COASTAL AND SHELF SEA MODELLING

THE KLUWER INTERNATIONAL SERIES

TOPICS IN
ENVIRONMENTAL FLUID MECHANICS

Series Editors

Dr. Philip Chatwin, *University of Sheffield, UK*
Dr. Gedeon Dagan, *Tel Aviv University, ISRAEL*
Dr. John List, *California Institute of Technology, USA*
Dr. Chiang Mei, *Massachusetts Institute of Technology, USA*
Dr. Stuart Savage, *McGill University, CANADA*

Topics for the new series include, but are not limited to:

♦ Small-to medium scale atmospheric dynamics: turbulence, convection,dispersion, aerosols, buoyant plumes, air pollution over cities

♦ Coastal oceanography: air-sea interaction, wave climate, wave interaction with tides, current structures and coastlines, sediment transport and shoreline evolution

♦ Estuary dynamics: sediment transport, cohesive sediments, density stratification, salinity intrusion, thermal pollution, dispersion, fluid-mud dynamics, and the effects of flow on the transport of toxic wastes

♦ Physical limnology: internal seiches, sediment resuspensions, nutrient distribution, and wind-induced currents

♦ Subsurface flow and transport (the unsaturated zone and groundwater): diffusion and dispersion of solutes, fingering, macropore flow, reactive solutes, motion of organics and non-aqueous liquids, volatilization, microbial effects on organics, density effects, colloids motion and effect, and effects of field scale heterogeneity

♦ Debris flows, initiated by lava flow from volcanic eruptions; mud flows caused by mountain storms; snow avalanches, granular flows, and evolution of deserts

♦ Oil spills on the sea surface and clean-up

♦ Indoor contamination: transport of particles in enclosed space, clean room technology, effects of temperature variation

♦ Risk assessment: industrial accidents resulting in the release of toxic or flammable gasses, assessment of air and water quality

♦ New methods of data acquisition: the use of HF radar, satellites, and Earth Observation Science

♦ Stochastic models and Mass transfer

COASTAL AND SHELF SEA MODELLING

by

Philip Dyke
University of Plymouth, United Kingdom

KLUWER ACADEMIC PUBLISHERS
Boston / Dordrecht / London

Distributors for North, Central and South America:
Kluwer Academic Publishers
101 Philip Drive
Assinippi Park
Norwell, Massachusetts 02061 USA
Telephone (781) 871-6600
Fax (781) 681-9045
E-Mail <kluwer@wkap.com>

Distributors for all other countries:
Kluwer Academic Publishers Group
Distribution Centre
Post Office Box 322
3300 AH Dordrecht, THE NETHERLANDS
Telephone 31 78 6392 392
Fax 31 78 6546 474
E-Mail <services@wkap.nl>

   Electronic Services <http://www.wkap.nl>

Library of Congress Cataloging-in-Publication Data

Dyke, P.P.G.
    Coastal and shelf sea modelling / by Philip Dyke.
     p. cm. -- (Topics in environmental fluid mechanics ; 2)
    Includes bibliographical references and index.

    1. Oceanography--Mathematical models. I. Title. II. Topics in environmental fluid
    mechanics ; EFMS2.

GC10.4.M36 D93 2000
551.45'7'0113--dc21

                                                                00-064075

ISBN 978-1-4419-5013-0

Printed on acid-free paper.   Printed in the United States of America

The Publisher offers discounts on this book for course use and bulk purchases.
For further information, send email to <molly.taylor@wkap.com> .

To Eleanor

Contents

CONTENTS

Preface

For the past ten years or more, the author has been responsible for the delivery of a modelling module on a graduate studies degree in marine science. It became apparent that there was no suitable textbook to cover the material, so this text is an attempt to fill the gap. The assumption initially is that students will have no more than a very rudimentary knowledge of mathematics. The MSc in marine science attracts students with a wide variety of undergraduate degrees as background. Biologists and geologists in particular who have not traditionally been exposed to much calculus. An earlier book Dyke (1996) indeed attempted to do modelling with no mathematics at all and with partial success. On the other hand, the amount of modelling possible with little or no mathematics is rather limited. In this text, a compromise is reached. There are large sections where there are few equations, yet there are other places where equations abound. Indeed, it may seem in places that it has been written by two people; not so, I wrote it all. If you have a background that can cope with a level of calculus normally found in an undergraduate degree in a physical science or engineering, then all of the book should be readily accessible. Others will need to tread more carefully, but there are plenty of words between the symbols that describe what is happening.

The motivation for studying how to model coastal sea processes is by and large environmental protection. The sea is being subject to discharges and pollution by man as well as climate changes perhaps also due to mankind. Understanding how these changes in input effect currents, water quality and sea levels is important and understanding how these are modelled the main purpose of this book. The first chapter sets the scene, outlining this motivation and explaining the modelling process in general. The second chapter overviews dimensional analysis which is a very useful modelling tool and goes on to outline basic dynamic balances in the sea. In order to get the most out of this chapter, some knowledge of calculus is necessary. The third chapter is a brief account of the numerical methods used to solve the modelling equations. It has been written in such a way as to minimise the amount of technicalities instead concentrating on giving an overview of the various methods, their consequences and an account of possible errors and inaccuracies. The fourth chapter is on the all important subject of boundary conditions; what they are and how they are represented mathematically. The first four chapters contain all that is required in terms of techniques. In chapters five, six and seven separate areas for mod-

modelling are described. Chapter five for the vital subject of diffusion, chapter six for continental shelf sea modelling including tides and storm surges and chapter seven for other environmental impact such as ecosystems modelling and coastal erosion studies. In chapter seven there is also a necessarily very brief account of waves and wave prediction. This is there for completeness but will mainly interest civil engineers. Finally, in chapter eight there is a chance to do some modelling with the author acting as tutor. Some background statistics is given, then there are a series of examples based on the contents of the first seven chapters. It is recognised that there are considerable differences in level in these examples. Please choose those most appropriate and as for the rest, either scorn as trivial or ignore as unsuitable. The answers to all the examples and exercises are there, but the reader is given the opportunity to work through each example first. The book is not written to be read from cover to cover like a novel. There is some repetition as is inevitable (ocean modelling is not a linear science) and cross referencing is used to flag this. The notation is as consistent as possible, and in the very few places where there are clashes these are explained and should not confuse.

It is a pleasure to acknowledge all my students over the years for the feedback on the lecture material which has helped a great deal in the selection of the subject matter for this book. Thanks also to the participants at the biennial JONSMOD (JOint Numerical Sea MODelling Group) who have done much to keep me abreast of the latest developments. Finally, thanks to all those who have given permission to use diagrams from other publications. All of these have been dutifully acknowledged.

Phil Dyke
Professor of Applied Mathematics
University of Plymouth
July 2000.

Chapter 1

The Modelling Process and Environmental Impact

1.1 Introduction to modelling

To the majority, the word modelling still means something to do with photography or, if they have a scientific background, the building of scaled-down replicas that ought to mimic real life situations. In this latter category one thinks of Civil Engineering consultants building models of harbours with attendant breakwaters and jetties, and then subjecting them to a particular wave climate. The way this is done is to build a physical model, usually in a large area reminiscent of an aircraft hanger. In this model, the area of coast or river or estuary (whatever) is built from materials such as concrete, sand and cement. Of course there is a scale, perhaps 1:20 or even larger, which needs to be considered when examining results. If waves are of interest, then there has to be a paddle mechanism included in order to generate them. Exactly how the scale factors can be calculated is the subject of Chapter 2, but suffice it to say that measurements of quantities such as wave height, current speed and direction, the force on pier or jetty can be made on the model. Appropriate scale factors are then applied and an estimate of the real life wave height, current speed and direction, force on the pier or jetty or whatever can then be made. Up to thirty or so years ago virtually all modelling in coastal engineering or oceanography referred to this kind of activity. These days, modelling invariably means use of the computer and the big once national facilities (e.g. Hydraulics Research in the U.K. Delft Hydraulics in The Netherlands both now privatised) now have much scaled down (no pun intended) the facilities for these physical models but have many sophisticated computer models to replace them. Many would prefer the word enhance rather than replace, as there is still the place for a physical model where the carefully placed strain gauge can give information to reinforce the output from a mathematical model. In most cases, the results from a mathematical model implemented via software on a computer will tell the same story as the results from a physical model, but if there are contradictory results, neither should automatically be believed. Perhaps they are both wrong and the situation is more complicated that either model builder thought! Areas where physical modelling is still dominant is in the building of bridges and in the design of spacecraft. In both of these areas, the final costs are so huge that the expense of building a physical model is less critical than for example estimating dilution rates of a dissolved substance in an environmentally sensitive estuary. In this latter case, mathematical models are now almost always used.

We shall not be discussing physical models in this text. The parallels will be explored a little further in chapter 2 where this is natural, but thereafter physical models are left behind. Mathematics is often thought of as operating within a very well defined set of axioms using well defined techniques to give precise answers to well defined problems. Pure mathematicians hold this view. However, such a rigid structure is not well suited to contribute to the description of a practical science such as oceanography. Papers that are very mathematical which may be very interesting in their own right often have only a tenuous link with reality. At the other end of the spectrum there are some very simple mathematical models that embody the essence of oceanographic truth, and we will definitely be meeting some of these. It is this blending of mathematics with the knowledge of oceanographic processes in which the art of successful mathematical modelling lies.

There is no doubt whatever that mathematical modelling has been greatly assisted by the rapid advances in computer power. Nevertheless, mathematical modelling can still certainly take place without it, it is just that computing increases the scope of the modelling. The emphasis in this text is definitely not on computing which may be thought of as a tool that enables modelling to take place. Instead the concentration is on the correct mathematical description of the ocean physics. Oceanography and marine physical science may be thought of as synonymous. It is a science that has grown through painstaking observation and progressed through scientists making judgements and deductions from these observations. There are several distinctive features that although not peculiar to oceanography, meteorology and earth science in general make it particularly amenable to the relatively new art of mathematical modelling. (Yes, although there is a great deal of scientific method and rigour in mathematical modelling, it still remains in many ways very much an art!) First, oceanography as an applied science has to incorporate aspects of physics, chemistry and biology. Indeed it may be argued that the sea provides an ideal vehicle for the study of some (but certainly not all) of the fundamentals of these basic sciences. In order to understand some of the processes that go on in the sea, it is therefore necessary to simplify some aspects and ignore others. This is what occurs in modelling. Secondly there is a very important aspect to modelling called validation. In most sciences and engineering, validation means trying out the model and comparing it with the real situation. In oceanography, the entire history of the science from the accumulated wisdom of fishermen through the voyages of discovery to modern day scientific expeditions is centred around observations and provides in some respects an ideal scenario for validation. Conditions are however not controlled, as the day has not arrived where weather can be prescribed therefore there is no control over at least one input. This is certainly a disadvantage in some respects, although it does encourage the continuation of the lively debate between the modeller and the observer.

1.2 Environmental Issues

In the last few years many new substances have been developed, from those used in foodstuffs, packaging, building materials to paint additives and the many plastics used everywhere in our daily lives. With the industrial processes that are used to produce these substances has come an awareness that care has to be taken about the disposal of the byproducts of the manufacturing. The environmental lobby has become particularly strong in the last fifteen to twenty years, and it is really only now that we are at last beginning to become aware of the lasting effect of all the foreign material mankind seems to be continually pumping into the earth's environment. Of course, much research needs to be done before our understanding is anywhere near adequate, but now whenever there is a new manufactured chemical or the production of a hitherto unsuspected byproduct there is in most countries strict legislation governing what can and cannot be allowed. Unquestionably, sometimes the environmental lobby prevents what is an innocent process taking place, but this is much less worrying than permitting a pollution that may unwittingly cause widespread environmental damage.

Some worrying case histories have come to light following the collapse of the Soviet Union. In Poland, Czechoslovakia (now split into the Czech Republic and Slovakia) and other former Warsaw Pact countries cumulative environmental effects have shortened the life expectancy of people living within the range of rivers polluted by the unthinking discharge of industrial waste including that from nuclear industries. Major rivers in Poland were so polluted that they were useless even for industry let alone recreational use. Using them for drinking water was completely out of the question. Before 1991 East Germany (as it was then called) was encouraged to increase industrial production regardless of any impact this might have on the environment. In December 1990, drinking water in Brandenberg did not meet EU (European Union) quality standards. Tankers had to be used to supply drinking water in some areas. Nitrate levels were up to 25 times EU legal limits. Samples from the Havel river contained high levels of phosphate, ammonium, benzol and zinc. Purification ensued, but even then water was found still to be contaminated with oil, even phenol. Post 1991 the west has become horrifically aware of dead rivers and dead lakes with toxic levels of similar chemicals. The unified Germany is doing its best to clean up the mess left by its communist predecessor, but it takes both time and money. It is a cruel irony that Germany now comprises two pre 1991 countries that were at the opposite ends of the environmental spectrum as far as cleanliness is concerned. A cruelty that has hit the ex West German where it hurts; in his pocket! Some of these more legal and political issues are outlined in section 1.5.

There are many ways in which chemicals can effect the environment. Most of these are local effects and come under the name pollution. There are two effects however that have had an impact over the whole globe since the mid 1980s. In order to reach the public attention these days requires the adoption of a "sound bite" (itself a sound bite incidentally; interesting for those aware of Russell's Paradox!). The sound bites in question are "global warming" and "ozone depletion". If the public can get things wrong it usually does, and these effects although very independent have been confused. To sum up global warming, this is the increase of (man made?) gaseous discharge of chemicals such as carbon dioxide and sulphurous oxides into the atmosphere which alters the balance between incoming and outgoing radiation, decreasing the latter so that the earth as a whole warms up. That the earth's albedo (as this balance between incoming and outgoing radiation is called) is changing is beyond question but whether it is through industrialisation or the wholesale eradication of equatorial rain forests (or some of both) is still controversial.

There is less controversy about ozone depletion. This is the name given to the disappearance of that layer enveloping the earth largely comprising ozone that is responsible for shielding us from the harmful effects of the sun's ultra-violet rays. The aerosol can is less than 100 years old, and in the last twenty years it has mushroomed in use until the gas used for the propulsion of the spray (CFC or Chlorofluorocarbons) was found to persist long enough to attack and destroy parts of the ozone found in the upper parts of the atmosphere. As has been said briefly above, this ozone protects us from the ultra-violet rays arising out of the sun and overexposure to which can lead to skin cancers. This whole ozone depletion problem has been brought to the attention of the public by the publication of colourful pictures of the increasingly large hole in the ozone layer over the antarctic, and a smaller one discovered over the arctic. The problem in fact has been known for some time; the depletion of stratospheric ozone was first observed as long ago as 1979, but not reported until 1985. A hole was observed in the ozone layer which is largest in the springtime, but at this time no culprit was identified. It was perhaps an oddity that would go away. Not so. By 1988, it was recognised that CFCs had an important role and the Montreal Protocol was signed: 100 countries signed a protocol in 1988 to discuss ways of arresting the depletion of ozone in the upper layers of the atmosphere. By 1990 and 1991 there were two successive years of severe ozone depletion, it was not going away and CFCs were confirmed as being linked to the problem. Although CFCs, first introduced via aerosols in the 1930s were contributing directly to ozone depletion, there was additionally a positive feedback mechanism at work. The CFC derived chlorine lowered air temperatures which in itself made CFC derived chlorine more effective in depleting ozone. By 1993, ozone had depleted to 21% of "normal" values. In 1996, the hole was as big as USA and Canada combined and by 1998 it extended over an area nearly twice the size of the antarctic continent itself. Not until 2015 will there be any sign of recovery. The greenhouse effect interacts with this ozone depletion problem by cooling the upper atmosphere even though the lower atmosphere is warming, which leads to the maintenance of the hole. All this is a salutary lesson in unforeseen environmental consequences to man made chemicals. Besides UV rays (leading to the enhanced risk of skin cancer as well as other health problems) other consequences, as yet unknown, may still await us. Not before time, in December 1995 signatories to the Montreal Protocol met in Vienna and agreed limits on ozone depleting substances (methyl bromide and LDCs, the former to be phased out by 2010, the latter to stabilise at 1995-1998 levels by 2002).

These global problems have helped and continue to help to focus public attention on the environment and how important it is to protect it even though these two large problems are perhaps beyond the individual to influence to any measurable extent. Nevertheless, people are now aware that they must "do their bit" to protect the environment whether this is by being careful about waste disposal (who had heard of a bottle bank thirty tears ago?), or by using their cars less. Nowadays, using unleaded fuel in the family car or supporting so called organically grown foodstuffs is recognised good practice. This text will only touch on such difficult world wide environmental problems, instead the focus will be on smaller scale modelling, and these large problems form the context in which many of the smaller models are embedded. In the next section let us look at the modelling process itself.

1.3 The Modelling Process

In many books on mathematical modelling, the starting place is the description of some kind of idealised modelling process using as a vehicle some equally idealised problem. The trouble with this is that both students and experienced practitioners alike find this less than convincing. The element of trial and error that seems to be involved in the classical modelling process is unrealistic to the practicing oceanographer, whilst to students of ocean science the whole process looks too ideal, not related to actuality. However it is the heuristic trial and error side of modelling that makes it so successful in its mimickry of real life. With this in mind we design our modelling process particularly with the ocean scientist in mind.

The singular most important aspect of modelling that has led to its recent popularity is the ready availability of cheap but increasingly powerful computers. In the whole of the 1960s and 1970s and the first half of the 1980s in order to use these computers it was necessary to be able to programme them. In order to programme them it was necessary to learn a high level computer programming language such as ALGOL (in the early days), FORTRAN, PASCAL or C. The details of the programming in turn demand a detailed knowledge of mathematics and the numerical methods that are used to translate the mathematics into the discrete mathematics that computers can use. By their very nature, these programs utilising as they do powerful computers are complicated and are based on sophisticated rather than simple mathematics. A requirement for those involved in marine modelling was therefore some knowledge of mathematics including the calculus that is used to describe the dynamic balances in a fluid and the transport of heat and salt, and the methods of discretising that in turn demand knowledge of numerical methods. Much of this kind of modelling is still of course going on, but unquestionably it is no longer mandatory to be as close to the mathematics. Many marine scientists are concerned with models on computers because they wish to answer engineering or environmental impact questions, but they lack the mathematical background to formulate and then program models on computers themselves. Software (the modern name for a computer programme that is commercially available) is only a successful product if it accessible to the majority of likely users. Of course it must also be useful in terms of producing meaningful results. It is widely recognised that not enough marine scientists have the mathematical background to comprehend the details of today's marine models, and it is indeed fortunate that this is no longer necessary. The very computer power that enables the models themselves to be complex also enables so-called "front ends" to be incorporated into models. These front ends act as an interface between the program and the lay user and enable the lay user to use the program constructively without the need to get involved in the programming itself. One common method is to use English commands to enquire of the user what features he or she wishes to incorporate into a model. In effect the user operates with a series of menus, choosing from a set or answering yes or no to simple questions. In this way, a particular problem can be solved by adapting a complex program through menus and without detailed programming knowledge. This is the general philosophy behind expert systems that are now quite widespread especially in the medical field.

The general philosophy behind modelling in marine systems can be expressed succinctly in a flow chart of the type shown in Figure 1.1. In the language of systems, ocean science might be thought of as a mixture of soft system and hard system. A soft system is one that is ill-posed and usually involves humans, whereas a hard system is one that is controllable, obeys well formed laws and is by and large amenable to exact mathematical solution. Perhaps a more natural division for marine science is into three classes; natural systems (e.g. biological organisms); artificially designed but physical systems (e.g. engineering devices); and artificially designed but abstract systems (e.g. economic models and models involving scheduling). The kind of modelling indicated by the flow chart of Figure 1.1 mostly fits into the first of these categories, however models of the physics of the sea are very different from biological models. Most models of ocean physics arise from the application of well established laws and lead to well posed mathematical problems. Therefore in systems methodology they are hard systems. Biological models related to the ocean are also in fact hard as although there are no universally recognised laws with the same stature as physical laws, they are well posed and solutions exist. It is only very recently that ocean scientists have brushed with what might be termed soft systems. They are only soft because the sheer complexity of the latest models renders exact predictability difficult. With the development of genetic algorithms the distinction between hard and soft systems is blurring. Related neural network modelling seeks to simulate systems by emulating the way humans tackle problems, through training and learning. These notions are in their infancy and have not yet reached the stage where they can be applied systematically to environmental problems.

Thinking of modelling itself, it is perhaps tempting to describe it as a well

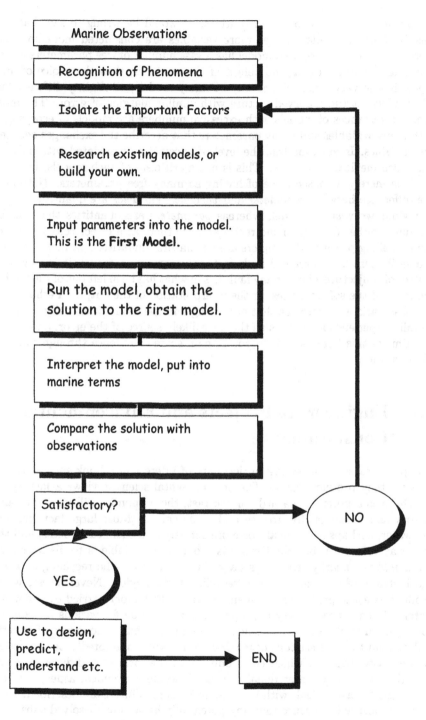

Figure 1.1: A marine modelling process flow chart

established set of principles and routines. It would be wrong to say that this was far from the truth, but as more variables are considered modelling gets more of an art. If a proposed model has a vast number of free parameters whose values are more or less at the behest of the modeller, then the choice of final model is also very large. There may be many models, all slightly different but all of which fit the observed picture with similar margins of error. The most common instances of models with excessive numbers are models of ecosystems with many variables and many parameters that describe the exchanges between the variables. In such models, the level of uncertainty associated with outputs must also inevitably increase. This is not a criticism of such models, far from it. It is merely a consequence of having so many free parameters. Note here the difference between *parameter* and *variable*. *Variables* are quantities whose behaviour we wish to model, whereas *parameters* are quantities that can be estimated either directly or indirectly and are there as a direct consequence of the modelling process. Parameters are not natural quantities, variables are. Of course if there is a perceived fault in the outcome of a model, it is usually a matter of conjecture whether any fault lies in what has been left out of the model, the method the software uses, or the interpretation of the output. Perhaps it is the observations themselves that are wrong and the model is actually working. Usually experience alone tells us the most likely source of the error, and one of the aims of this book is to help you to gain some experience through the eyes of the author.

1.4   Engineering Projects and Environmental Consultancy

The prime reason engineering projects are of interest in a book such as this is because they produce waste. The environmental scientist views a factory on the shore as a potential hazard. In contrast, the economist will view the same factory as a potential for growth and employment. Most large factories will use water, perhaps as a coolant or more actively in a chemical process, and the only way to ensure that the factory is economically viable is to discharge the water into the nearby river or estuary. In recent times, the recycling of waste has increased due to advances in recycling technologies. Nevertheless, waste products are still produced which emerge from the factory carried by the waste water. When a new factory is proposed, it is common for various parts of its construction to be put out to tender. This means that various sets of experts will be paid to build sections of it. One section will undoubtedly be the design of the waste disposal, which commonly takes the form of a diffuser to make sure that any waste water remains legal. The more technical aspects of how discharges behave is dealt with in Chapter 5 where diffusers are described. Here we are concerned to ensure that any potentially hazardous dissolved substance present in the waste may remain within environmentally acceptable values as laid down by legislature. In order to do this, the team involved in designing the diffuser often contains someone who is capable of building (or getting hold of) an appropriate mathematical model. This model should be capable of simulating the worst case scenario whereby the levels of the hazard are in some sense maximal. If even this maximum level is legally acceptable, then normally the go ahead for the manufacture of the diffuser as designed can be given. Some of the more technical aspects of flumes are covered in section 5.4.4.

The environmental questions that arise from not only the building of factories but many other activities such as sport, the modification of harbours, housing developments etc. has meant that environmental consultancy is now big business. As the population increases, and man explores and exploits more of the earth's surface, the environmental scientist assumes the role of a guardian. It is now certainly not optional to consider most carefully environmental questions every time industrial or other unnatural activity impacts on the world around us. Some of the things that were done in the name of industrial advancement in the UK are now seen as very wrong and make us wince. Unfortunately, like activities are still going on in other less civilised parts of the world. The framework for the control of this is the law which forms part of the next section.

1.5 Legal Issues and Environmental Protection

When man entered the industrial age back in the early 19th century, he started releasing chemicals into the environment. However back then the quantities were not large and the sociology and knowledge of environmental chemistry at this time also meant that any discharges were disregarded. More recently, things have certainly changed. The accidental or deliberate release of chemicals into the environment is now a very live issue. When a chemical is so released a whole host of processes ensue. As far as the activity of man is concerned, there are legal processes which will be outlined later. To give a flavour of the scientific processes, first of all the introduced chemical will interact with naturally occurring chemicals in the environment. This interaction could be just mixing, but could also involve chemical reactions with important consequences for the environment. Then there is the physics of how the introduced substance interacts with the environment. In water (river, estuary or sea) this could involve diffusion, sedimentation, or simply the migration of the substance with local currents. In addition the substance could be buoyant and contribute to a surface slick or sink and interact with the sediment. Finally the introduced chemical could interact with the biology; the plankton, fish, aquatic reptiles, amphibians, mammals and ultimately man. It is this interaction, almost always the most difficult to forecast that of course is the most worrying for the public and lies behind the heightened interest in environmental protection. We have to be so very careful as biological consequences are the most difficult of all to predict. To cite a non marine example, who would have thought that an insecticide (DDT) would cause birds that accidentally (incidentally really as they were more interested in the insects) consumed it to lay eggs that had abnormally thin shells. These shells could not withstand the weight of the incubating adult, hence the population of birds (peregrine falcons actually) was decimated. In the medical field we all now know the consequences of a pregnant woman taking the anti morning sickness drug thalidomide. It is this kind of worry that is behind the (over?) reaction of the general public (or do I mean the popular UK press) to genetically modified crops. A public, it must be remembered which is still smarting and facing uncertainty over BSE (Bovine Spongiform Encephalopathy) and its relationship to new variant CJD (Creutzfeldt Jakob Disease).

In recent times, the control and management of any waste products of industry have become increasingly governed by legislation. Pollution laws as they are colloquially known tend to be different in different regions of the world. In the USA the laws are usually very strict in terms of health, but less so in environmental protection terms. In Singapore all environmental laws are very strict indeed. In the old eastern block countries (notoriously East Germany as it tried to build its industry up from the ravages of the Second World War) environmental legislation was virtually non-existent or at least ignored even when it was there. Even in the UK pollution laws can be different in England and Wales than in Scotland and Northern Ireland. Legal principles are there to protect the interests of those who can be threatened or damaged by pollution. The legal process is notoriously expensive and often tortuous, therefore it is in everyone's interest (apart perhaps from the lawyers) not to go to court. Most cases are thus settled by insurers. Some large cases however need a national forum and usually arise because they are in some way new. Obvious examples are the large tanker accidents the Torrey Canyon (1967) and the Amoco Cadiz (1978) and more recently the Exxon Valdez where liability was contested, not surprisingly given the extent of the environmental disaster in each case. On a smaller scale, the liability for minor slicks caused by the flushing out of "empty" tanks lies squarely with the captain and the only problem is catching him (or her!). More relevant to modelling is the legislation that exists to prevent more than particular concentrations of certain chemicals in the sea. This is the maritime equivalent to monitoring lead in car emissions (which led to the development of lead free petrol and the catalytic converter). Examples of what is meant in the maritime context is the pollution caused by painting boats with anti fungal paint (tributyl-tin), or the control of heavy metals that arise from chemical process factories (Lead (again), Mercury and Cadmium), enhanced levels of which can occur in the waste water which is discharged into the nearby river or estuary. Experts usually formulate levels of chemicals that are deemed reasonable to tolerate, and if these are exceeded prosecutions ensue. As new processes are developed, these are scrutinised and the legislation is modified accordingly. The enforcers of the legislation vary. Sometimes it is a national body such as the River Authority (in England and Wales) or the River Purification Boards (in Scotland), sometimes the transgressor is in breach of a law such as the Environmental Protection Act in which case the police can be involved. Successful prosecution can result in the closing of a factory and the fining or imprisonment of offenders. As authorities become convinced that industrial concerns can consistently meet minimum standards, they are able to issue some kind of licensing agreement. This grants the licensee to manufacture.

In the international sphere, there are international treaties that all signatories obey. Examples of this include the treaty that preserves Antarctica for scientific study, and the Treaty of Rome which set up the fundamental legislature for the European Union. In recent times the law has become more complex in Europe as EU legislation increases. In the maritime field there have always been complications due to offences taking place in international waters, or there being at least three nations involved. One which owns the offending material (usually oil), one which owns the container (oil tanker), or the country of origin of the captain and the country to which the territorial waters in which the act occurred belongs. To this complex picture needs to be added the question of market forces. It is still broadly true that the small guy fighting the multi-million pound company loses.

In the UK case law still plays an important part. That is the court is a very powerful body, and if a case sets a precedent perhaps in the level of compensation paid by a company that has caused bodily harm (for example) this is cited in similar future cases. Some kind of convergence then occurs if there enough cases. This is the power of "common law".

The big question of global warming is causing meetings with high profile politicians (Kyoto 1997, Rio 1998) but as yet no treaty worth the name has emerged due to the partisan interests of some of the delegates, contradictory differences between environmental desires and local political reality are irreconcilable. Recently, the representatives of the Florida Keys community are joining forces with other island communities around the world (Fiji, Samoa etc.) in the hope of providing a convincing lobby to the USA industries who are widely recognised as main contributors to the industrial gases that cause global warming. If the sea level rises continue, then these island communities will be no more in a century or two! This author wishes them all the best.

Very recently the UK instigated a Royal Commission on Environmental Pollution, and has issued very widely the findings. Almost all the emphasis of the report is on global warming, and some quite alarming statistics have been presented. As an example, Figure 1.2 gives a graph showing how the energy use has risen through the last century. The clear message is again how mankind has to reduce the amount of fossil fuels being burnt or face dire consequences. Given the very long time scales, the signs that mankind can reverse the effects of global warming, or the effects of ozone depletion (for which there is more agreement) are not good. There are some very recent encouraging signs, but not enough. One convincing strategy that will help is to provide accurate models, and to do this the underlying processes have to be understood. This is what this book is about.

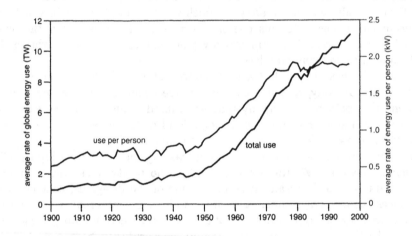

Figure 1.2: Growth in global energy use, 1900 - 1997

Chapter 2

Dynamic Balances

2.1   Introduction

In this chapter, we discuss notions of *typical* lengths, *typical* times, *typical* speeds etc. It is by no means obvious what is meant by this, therefore a few words by way of introduction are appropriate. Many, not to say most variables can be expressed in terms of time (measured in seconds), length (measured in metres) and mass (measured in kilograms). Examples are speed (length divided by time), density (mass divided by volume, hence mass divided by length^3) and pressure (mass divided by (length times time^2)). The quantities mass, length and time are called the *fundamental* quantities. In a given situation, most users of models as well as the builders of the models themselves wish to concentrate on phenomena which have a specific limited range of any of the fundamental quantities, or of variables derived from them. For example, a tidal modeller would not be interested in small time scales of a few seconds or large time scales of many centuries, but something in between. An estuarine modeller would not be concerned with length scales of hundreds or thousands of kilometres but something much less. Most phenomena in ocean science are scale specific due principally to the dominance of the Coriolis effect, and the art of successful modelling is more often than not linked to the ability to screen out the unwanted in order to focus on what is desired. Dimensional analysis forms an essential element of this simplification. Once this simplification has taken place, the terms that remain constitute a simplified dynamic balance and these balances are useful in describing the fundamental movements of the sea.

2.2   Dimensional quantities in Ocean Science

It has now been over 300 years since Sir Isaac Newton formulated his famous view of mechanics. Despite the recent advances in physics which have led to new theories of the very large and fast (general relativity, *circa* 1915) and the very small and also fast (quantum mechanics, *circa* 1926), the motion of virtually

all known objects closely follows the laws set down by Newton all those years
ago. Relativity and quantum mechanics are in fact not only incompatible with
Newtonian mechanics, they are at odds with each other! Fortunately neither
are relevant to ocean science which deals with less extreme mass, length and
time scales. In marine science, Newton's laws hold. The adaptation of Newton's
laws, notably his second, to fluids was done during the eighteenth century by
Euler and (Daniel) Bernoulli. In words, this law is "force = mass × acceleration" which in fluids reads "force per unit mass = acceleration". Bernoulli's
contribution was to convert the equation of fluid motion into the form of an
equation which expresses the conservation of mechanical energy.

The next logical step would be to derive these equations in their mathematical glory. We shall avoid doing this as this is not a mathematics text. Instead
we shall first discuss Newton's second law for fluid motion, often called the
Navier-Stokes' equation in terms of dimensions. In order to do this, all that
is required is some knowledge of what the balances represent and the magnitudes of fundamental quantities. As a prelude, let us generate a table of marine
quantities and their dimensions. Afterwards we will be little more mathematical.

Quantity	Unit	Dimensions
Mass	Kilogram(kg)	M
Length	Metre(m)	L
Time	Second(s)	T
Temperature	Kelvin($^{\circ}K$)	Dimensionless
Velocity	Metres per second (ms^{-1})	LT^{-1}
Acceleration	Metres per second per second (ms^{-2})	LT^{-2}
Area	Square metre (m^2)	L^2
Volume	Cubic metre (m^3)	L^3
Discharge	Cubic metre per second ($m^3 s^{-1}$)	$L^3 T^{-1}$
Force	Newton (N)	MLT^{-2}
Pressure	Pascal (Pa)	$ML^{-1}T^{-2}$
Pressure gradient	Pascals per metre (Pam^{-1})	$ML^{-2}T^{-2}$
Density	Kilograms per cubic metre (kgm^{-3})	ML^{-3}
Dynamic viscosity	Newton second per square metre (Nsm^{-3})	$ML^{-1}T^{-1}$
Kinematic viscosity	Square metres per second ($m^2 s^{-1}$)	$L^2 T^{-1}$
Surface tension	Newtons per metre (Nm^{-1})	MT^{-2}
Weight (same as force)	Newton (N)	MLT^{-2}
Angular velocity	Radians per second (s^{-1})	T^{-1}
Angular acceleration	Radians per second square (s^{-2})	T^{-2}
Vorticity	Radians per second (s^{-1})	T^{-1}
Circulation	Square metre per second ($m^2 s^{-1}$)	$L^2 T^{-1}$
Energy	Joule (J)	$ML^2 T^{-2}$
Work (same as energy)	Joule (J)	$ML^2 T^{-2}$
Power	Watt (W)	$ML^2 T^{-3}$
Temperature gradient	Degrees per metre ($^{\circ}Km^{-1}$)	L^{-1}

Table 2.1 *A Table of Parameters and their Dimensions*

In terms of dimensions, Newton's second law holds true since acceleration has dimension LT^{-2} and force per unit mass has dimension $MLT^{-2} \times M^{-1}$ which is also LT^{-2}. There are three different kinds of acceleration that occur in marine science (oceanography or meteorology). The first is straightforward and is the fluid acceleration; the exact counterpart of acceleration of particles in mechanics. To distinguish this from the others it is sometimes called point acceleration. There is no doubt that this has dimension LT^{-2}. The second kind of acceleration occurs because the earth is rotating. When Newton formulated his second law, it had to be in what he called an *inertial* frame of reference. This means that the origin of co-ordinates and the axes are fixed or at most travel with constant speed in a straight line. On the earth this is not so, and so Newton's second law has to be modified. Perhaps the best simple illustration of this is as follows. You can try this if you are lucky enough to have access to an old fashioned turntable of the type that plays vinyl discs, but if it is not yours, check with the owner first! Spin the turntable and lightly draw a straight line across the turntable without altering the speed of the turntable itself. Stop the turntable and you will see that the line is in fact curved. Now if the turntable is scaled up so that human dimensions are such that we are not aware of anything outside the turntable, then the line is curved seemingly for no reason. The way this is tackled is to invent an acceleration that acts on moving objects, moving them off the straight. (The line is actually the arc of a circle if the speed of turntable and chalk are both constant.) To return to the actual earth, if an iceberg is floating in a boundless sea and is given a push with no friction or any other forces acting, then the iceberg would not travel in a straight line but in a circle the radius of which is this speed divided by the local value of the Coriolis parameter (see Chapter 5 where the details of this are given). The Coriolis parameter is $2\Omega \sin(latitude)$ where Ω is the angular velocity of the earth $(= 7.29 \times 10^{-5}s^{-1})$ and is usually denoted by the letter f (or γ if f has to be used elsewhere). The acceleration caused by the rotation of the earth affecting the speed of an ocean current is $f \times$ current. This has dimensions $T^{-1} \times LT^{-1}$ which is once again LT^{-2}, i.e acceleration. *Coriolis* acceleration as this is called turns out to be the most important acceleration for oceanographers and meteorologists. Acceleration can also be formed by the combination $U^2 L^{-1}$ where U is speed. Acceleration formed this way does not contain time explicitly, it is called *advective* acceleration and can be present even when the current is steady. A river with a steady flow rounding a bend is quite a good illustration of advective acceleration. As the water rounds the bend, adjacent water particles although travelling with constant speed and therefore not possessing point acceleration change position relative to one another. Therefore there is a change with time in some respect even though the flow is a steady one. This acceleration is in fact exactly the same as the acceleration experienced by someone going around a corner. It is erroneously called centrifugal force and more properly called centripetal acceleration. Centripetal acceleration is u^2/r where u is speed and r is the radius of curvature of the bend. This is obviously of the form $U^2 L^{-1}$ mentioned earlier. We have thus talked our way through the three kinds of acceleration. In mathematical terms for the point acceleration,

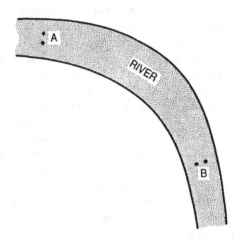

Figure 2.1: Adjacent particles flowing down a river

advective acceleration and Coriolis acceleration they take the form:

$$\frac{\partial \mathbf{u}}{\partial t}, \quad (\mathbf{u}.\nabla)\mathbf{u} \text{ and } 2\mathbf{\Omega} \times \mathbf{u}$$

respectively. The first, the point acceleration looks enough like the rate of change of **u** with respect to time for those familiar with elementary calculus to be convinced. The second is the advective acceleration and is not straightforward to derive. The symbol ∇ is the vector form of a spatial rate of change. The term thus represents the acceleration of a fluid due to the changes in relative position of all the fluid particles that surround the one in question. The third term is the Coriolis acceleration and is so important that it will now be derived. However this derivation is quite mathematical in nature involving as it does vector differentiation and cross products so may be glossed over if necessary. It is the result that is important. The derivation that follows also demands elementary geometry and some knowledge of vector algebra. The arbitrary point on the earth's surface is labelled O and is the origin of the local (x, y, z) co-ordinates. x points east, y points north and z points up. In this co-ordinate system shown in Figure 2.2 the velocity of the fluid is **u**, but of course the origin is moving with respect to the centre of the earth. (The earth's centre will be assumed fixed as its motion around the sun has such a large radius of curvature that it is straight to a very good approximation.) The motion of the point O is by virtue of the angular velocity of the earth. In the plane of the line of latitude that contains O, the x and y axes rotate with angular velocity $\Omega \sin \theta$ where θ is latitude. This is the vertical (z) component of Ω. The vector Ω has direction south pole to north pole. Thus, when the derivative (rate of

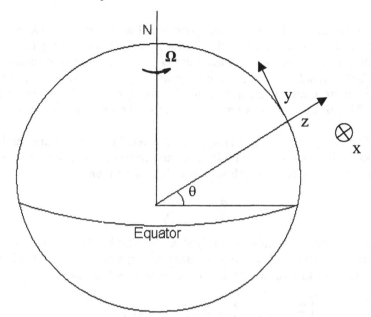

Figure 2.2: Motion relative to a rotating earth

change) with respect to time of any vector is calculated, and this calculation is performed relative to the rotating axes, the changes of the unit vectors that point along Ox and Oy, call them $\mathbf{i}$ and $\mathbf{j}$, need to be taken into account. So, take any vector $\mathbf{A}(t)$ which is a function of time then the calculation of its rate of change proceeds as follows:

$$\frac{d\mathbf{A}(t)}{dt} = \frac{d(A_1\mathbf{i} + A_2\mathbf{j})}{dt} = \frac{dA_1}{dt}\mathbf{i} + \frac{dA_2}{dt}\mathbf{j} + A_1\frac{d\mathbf{i}}{dt} + A_2\frac{d\mathbf{j}}{dt},$$

where A_1 and A_2 are the components of $\mathbf{A}(t)$ in the x and y directions respectively. By examining how the axes change in an infinitesimally small time, it can be easily shown that

$$\frac{d\mathbf{i}}{dt} = \Omega\mathbf{j}, \quad \text{and} \quad \frac{d\mathbf{j}}{dt} = -\Omega\mathbf{i}$$

and hence

$$\frac{d\mathbf{A}(t)}{dt} = \frac{dA_1}{dt}\mathbf{i} + \frac{dA_2}{dt}\mathbf{j} + \Omega A_1\mathbf{j} - \Omega A_2\mathbf{i}.$$

The first two terms are the rate of change of $\mathbf{A}(t)$ with respect to time as if the axes were not rotating. So the extra two terms represent the effects of the rotation. These can be put succinctly in terms of vector quantities by using the cross product. Thus we can write:

$$\left[\frac{d\mathbf{A}(t)}{dt}\right]_{fixed} = \left[\frac{d\mathbf{A}(t)}{dt}\right]_{rotating} + \boldsymbol{\Omega} \times \mathbf{A}(t).$$

The derivation of this has been two dimensional and may not look very general. However, it is mathematically rigorous as the above vector equation is co-ordinate independent, and a mathematical result that is co-ordinate independent but derived using specific co-ordinates (in our case two dimensional Cartesian co-ordinates) remains true in all orthogonal curvilinear co-ordinate systems. This is a useful theorem and illustrates the usefulness of pure mathematics!

The Coriolis acceleration is now easily derived by a double application of the formula. If r denotes the position vector relative to an inertial frame of reference (the centre of the earth for example) then we have

$$\frac{d^2\mathbf{r}}{dt^2} = \frac{d}{dt}\frac{d\mathbf{r}}{dt}$$

where all derivatives are relative to fixed axes. The left-hand side is the true acceleration. In terms of actually measurable quantities which are of course measured relative to a rotating frame of reference we thus have:

$$\frac{d}{dt}\frac{d\mathbf{r}(t)}{dt} = \left[\frac{d}{dt} + \mathbf{\Omega}\times\right]\left[\frac{d}{dt} + \mathbf{\Omega}\times\right]_{rotating} \mathbf{r}(t)$$

$$= \left[\frac{d^2\mathbf{r}(t)}{dt^2}\right]_{rotating} + 2\mathbf{\Omega}\times\left[\frac{d\mathbf{r}(t)}{dt}\right]_{rotating} + \mathbf{\Omega}\times(\mathbf{\Omega}\times\mathbf{r}(t)).$$

The Coriolis acceleration is the term

$$2\mathbf{\Omega}\times\left[\frac{d\mathbf{r}(t)}{dt}\right]_{rotating}$$

the term

$$\mathbf{\Omega}\times(\mathbf{\Omega}\times\mathbf{r}(t))$$

is the centripetal acceleration which is directed towards the axis of rotation of the earth but is so small compared to gravity that it is usually ignored. We have also assumed that the angular velocity of the earth $\mathbf{\Omega}$ does not vary with time. (Otherwise there would be yet another term to consider!)

With respect to the chosen co-ordinate system (see Figure 2.2) the Coriolis acceleration is

$$2\mathbf{\Omega}\times\mathbf{u} = 2(0, \Omega\cos\theta, \Omega\sin\theta)\times(u,v,w)$$

where $(u,v,w) = \mathbf{u}$ is the fluid velocity. The double use of the symbol u is less confusing than alternatives. u is the easterly current, v the northerly current and w the upward current. This latter is of course very small and is neglected except for very specialist modelling. w is always neglected when compared with u or v, and it is this approximation that gives the Coriolis acceleration the components $(-fv, fu, 0)$ where $f = 2\Omega\sin\theta$ and is called the Coriolis parameter. f is twice the vertical component of the earth's angular velocity, and it is only this component that plays an important part in the dynamics of the ocean and atmosphere.

Let us now turn to the side of the equation which contains the forces. There are three all very distinct forces that act on a volume of ocean or atmosphere. Think of this volume as in an arbitrary position but of unit mass. The first force considered is perhaps the easiest to understand in that we all experience it every day, it is the force of gravity. We have already dismissed centripetal acceleration as very small and negligible in comparison with gravity, in fact the measured gravity has to contain centripetal acceleration as the earth cannot be stopped from turning whilst experiments to determine g are carried out! Such gravity is called *apparent* gravity. To do the sums, true gravity is around $10ms^{-2}$ whereas the magnitude of centripetal acceleration is $\Omega^2 R$ where R is the radius of the earth. This is $(7.29 \times 10^{-5})^2 \times 6.36 \times 10^6$ in S I units. This is of order 3×10^{-3} certainly a lot less than g by a factor of over 1000. The second force is due to the surrounding fluid exerting a force on our chosen volume. This force is due to the difference in pressure across the volume. We call this the pressure gradient force as in the limit of very small control volume, it is indeed a gradient rather than a difference in pressure. Finally there are frictional effects also due to the surrounding fluid. In a laboratory and in laminar flow these are due to the viscosity of the fluid, at sea they are invariably due to eddy motion and are characterised as turbulence. The details of modelling turbulence come later in section 2.4.3. Dimensionally, "force per unit mass" has dimensions LT^{-2} which is the same as acceleration. Pressure gradient divided by density is $ML^{-2}T^{-2} \div ML^{-3}$ which is LT^{-2} as required. The dimensions of the frictional force due to turbulence is (see section 2.4.3) also LT^{-2}. In marine science there is usually an important difference between horizontal and vertical length scales. This difference is only absent in very small scale dynamics - sea surface waves or rivers for example. The different letter D (for depth) is used to denote typical vertical scales and because of the requirements of mass conservation, expressed mathematically by

$$\frac{\partial u}{\partial x} + \frac{\partial v}{\partial y} + \frac{\partial w}{\partial z} = 0,$$

it is essential that the vertical velocity has dimension UD/L or $U \times D/L$ where the dimensionless ratio D/L is called the *aspect ratio*. A similar term is used in aeronautical engineering.

2.3 The Buckingham Pi Theorem

In this section we shall do some calculations based on a knowledge of the dimensions of the quantities and not much else. Dimensional analysis can be usefully employed to determine likely relationships between variables or parameters, but first we have to know enough about the situation to state what the parameters or variable are and which can be ignored. If we get this wrong, dimensional analysis can be misleading which can be worse than useless. A useful tool in dimensional analysis goes by the rather grandeur name of the Buckingham Pi Theorem. It was developed by Lord Buckingham in 1915 and he used the symbol Π to denote products of quantities, so the name Buckingham Pi Theorem has stuck. It is now stated in its generality for reference, but it does not come alive until it is used.

Buckingham Pi Theorem

Given n parameters such as length, speed, density, viscosity, force etc. and given that these parameters are composed of a set of m quantities (in our case three; mass, length and time), then it is possible to express the relation between the parameters in terms of $n - m$ dimensionless products, formed from any n parameters regarded as primary.

To get to grips with what this actually implies, consider a non-marine related example.

Consider the problem of finding the period of swing of a simple pendulum. The pendulum consists of a string of length l which is fixed at one end. To the other end is attached a bob of mass m_0 and this bob is displaced a small distance from its vertical equilibrium position and swings freely. Now, we have to reason which parameters are important in determining the period of oscillation. Certainly the mass m_0 is one candidate, as is the length of the string l. One other parameter is the acceleration due to gravity g, this is because it is the component of the weight of the bob $m_0 g$ perpendicular to the string which is the restoring force. It is in this part of the process that errors are prone to be made; some insight into the working of the simple pendulum is mandatory to get the right parameters. The Buckingham Pi Theorem can now be invoked as follows: first of all $n = 4$ as we have parameters P, l, m_0, g (period, length of string, mass of bob and acceleration due to gravity respectively) and $m = 3$ (as our fundamental quantities are mass M, length L and time T). There is thus one $(4 - 3)$ dimensionless number. The equation is

$$P g^B m_0^C l^D = \text{const.}$$

The dimensions of the four parameters are

$$P \sim T, \quad g \sim LT^{-2}, \quad m_0 \sim M \text{ and } l \sim L$$

from which

$$T(LT^{-2})^B M^C L^D = \text{const.}$$

Equating powers of T, L and M to zero gives the equations

$$1 - 2B = 0, \quad B + D = 0, \quad C = 0$$

from which

$$B = 1/2, \quad C = 0, \quad D = -1/2.$$

This means that the quantity

$$P g^{1/2} l^{-1/2} = \text{const.}$$

Making the period P the subject of this formula gives

$$P = \text{const} \sqrt{\frac{l}{g}}.$$

The correct formula is obtained by setting the constant equal to 2π. There is no way dimensional analysis can give the value of this constant. It has however given the correct dependence of P on g and the length of the pendulum l, and it correctly predicts no dependence on the mass of the bob m_0. Thus had we forgotten about the mass of the bob, we would have still obtained the correct answer. This is down to luck; if we had forgotten g we would have obtained nonsense (try it!)

We now see how the theorem works. It gives a dimensionless grouping which can either be used to find how one of them depends on the others (as above) or it can be used to determine the number and form of dimensionless groupings. We shall now use the theorem for this purpose, this time via a marine related example.

2.3.1 Parameterising Bottom Friction

Friction or drag, caused by the action of the sea bed on the water just above it is a force. As such therefore it has dimensions MLT^{-2}. Physically one would expect this force to depend on the local speed of the water $U = LT^{-1}$. It must also vary with the typical length scale L as the larger this is, the greater will be the friction. The more viscous (or turbulent) the sea near the bed, the larger the drag force. Finally, the larger the density ρ, the larger the drag force. We shall be having more to say about how to parameterise turbulence in the next section, but calling it an enhanced viscosity will do for now and turns out not to be too wide of the mark. Let us now use the Buckingham Pi theorem and equate the frictional drag, F to powers of L, U, ρ and viscosity μ. That is, we write

$$F = L^A U^B \rho^C \mu^D.$$

Now, from Table 2.1 the dimensions of ρ are ML^{-3} and the dimensions of μ are $ML^{-1}T^{-1}$. With $F = MLT^{-2}$ and $U = LT^{-1}$ we obtain the dimensional equation

$$MLT^{-2} = L^A (LT^{-1})^B (ML^{-3})^C (ML^{-1}T^{-1})^D.$$

Equating powers of M, L and T on each side gives the three equations

$$
\begin{aligned}
M: & \quad 1 = C + D \\
L: & \quad 1 = A + B - 3C - D \\
T: & \quad -2 = -B - D.
\end{aligned}
$$

Of course these cannot be solved (three equations in four unknowns), but writing everything in terms of D we get

$$C = 1 - D, \quad B = 2 - D, \quad \text{and} \quad A = 2 - D$$

so that

$$F = \rho L^2 U^2 f\left(\frac{\mu}{\rho L U}\right),$$

where the last term on the right is an arbitrary function (the combination $\mu/\rho LU$ is dimensionless). Now, this is as far as we can go using dimensional analysis, but it turns out that there are good physical reasons wrapped around pressure being force per unit area for letting $L^2 f(\mu/\rho LU)$ be a constant for a given regime. This constant is often called C_D, especially in civil engineering. Hence we arrive at the expression

$$F = \rho C_D U^2,$$

which is the quadratic friction law. This law has been widely used by civil and coastal engineers concerned with the flow of water over a river, estuary or coastal sea bed for over a century, and is still in use today. If the current is a vector, then in order to maintain that the drag and current must be in opposite directions we write

$$\mathbf{F} = -\rho C_D \mathbf{u}|\mathbf{u}|.$$

This is the accepted form of the quadratic drag law.

We shall now move on to consider another role for dimensional analysis in oceanography. First however some specific balances need to be discussed.

2.4   Fundamental Balances

2.4.1   Geostrophy and Hydrostatic Balance

The dominant acceleration in the sea when length scales are greater than $10^4 m$ ($10km$) is the Coriolis acceleration which has horizontal components $(-fv, fu)$. This acceleration is balanced by the pressure gradient except near the surface and solid boundaries where friction is important. Vertically the pressure gradient is balanced by gravity. The only exception to this is for wind driven waves which we will not consider until Chapter 6. Writing p for pressure, ρ for density for the majority of the worlds oceans and seas therefore, the three equations

$$-fv = -\frac{1}{\rho}\frac{\partial p}{\partial x}$$

$$fu = -\frac{1}{\rho}\frac{\partial p}{\partial y}$$

$$\text{and} \quad 0 = -\frac{1}{\rho}\frac{\partial p}{\partial z} - g$$

indicate the overall dynamic balance. The first two of these equations are geostrophy (or the geostrophic equations) and the third is hydrostatic balance. Most of us are acquainted with hydrostatic balance; it states that the pressure anywhere in a fluid is solely due to the weight of water above it. Under a sea surface which is expressed by the equation $z = \eta(x, y, t)$ this equation integrates to

$$p(x, y, z, t) = p_A + \rho g(\eta(x, y, t) - z)$$

where p_A is atmospheric pressure and $z = 0$ is mean sea level. From this equation, differentiation immediately deduces the following equivalent expressions for geostrophic flow

$$-fv = -g\frac{\partial \eta}{\partial x}$$
$$fu = -g\frac{\partial \eta}{\partial y}.$$

Although this is very idealised, it is nevertheless of practical value. For example the original geostrophic equations in terms of pressure can be used to deduce the flow of air around a low pressure centre (anti-clockwise in the Northern Hemisphere, the reverse south of the equator). For an oceanographic application, look at the first of these equations. This implies that if v is large, then there is a sea surface slope ($\partial \eta / \partial x > 0$) in which the sea surface is high in the east, low in the west (north of the equator, the reverse south of it). To realise that a flow causes a sea surface slope at right angles to the flow is the crucial point here. In straits where the flow can be large (due to a tide perhaps) the difference in sea level measured across the strait can be substantial enough to surprise the unsuspecting. For example, with a current of $5 \ ms^{-1}$ through a strait of width $1 \ km$. at a latitude where $f = 1.2 \times 10^{-4} \ s^{-1}$ a difference in height of $6cm$. results from geostrophy and this can be easily measured.

Another simple balance can be derived quickly from geostrophic flow. The two equations that form this balance are called the *thermal wind equations* a phrase that arises from their meteorological origin. The hydrostatic balance equation can be written:

$$\frac{\partial p}{\partial z} = -\rho g.$$

Assuming that the density ρ depends on x and y, differentiating each geostrophic equation with respect to z and substituting for $\partial p / \partial z$ gives:

$$f\frac{\partial v}{\partial z} = -g\frac{\partial \rho}{\partial x}$$
$$f\frac{\partial u}{\partial z} = g\frac{\partial \rho}{\partial y}.$$

The implication of these equations is that a horizontal density gradient (salinity or temperature usually) induces a change in velocity with depth. More importantly perhaps, if both u and v are independent of depth, then there can be no horizontal density gradients. In the atmosphere, the presence of horizontal temperature gradients can produce updrafts which are particularly sought after by glider pilots. The thermal wind equations neatly express this. Let us examine further some balances and dimensionless numbers associated with density gradients.

2.4.2   Internal Structure

Up to now, dimensional analysis has been used to deduce fundamental balances that are valid over the entire width of a basin, estuary or coastal sea. Some are even valid over the entire ocean. The business of taking characteristic length and time scales seems to imply this. However, there is an internal structure to the sea in which density changes play an important part. For example, it is well known that most oceans contain a pycnocline which is a sharp change in density. This usually marks the depth at which the surface mixed layer meets the deeper cooler water which is unaffected by the wind. In some places the pycnocline is a thermocline (sharp change in temperature) as in most seas and oceans, and in some places it is a halocline (sharp change in salinity with depth) as in some estuaries. A useful quantity which is linked to change in density with depth is the ratio

$$\frac{\text{gradient of density}}{\text{ambient density}}$$

which has dimension L^{-1}. If this quantity is multiplied by g the acceleration due to gravity, one has a quantity which has the dimension T^{-2} or the square of frequency. It is usually written

$$N^2 = -\frac{g}{\rho}\frac{\partial\rho}{\partial z}$$

the negative sign being there to indicate stability as density has to increase with depth otherwise the sea overturns (recall z points upward). N is called the buoyancy frequency or, less illuminating, the Brunt-Väisälä frequency. The buoyancy frequency can be realised as follows. Imagine a density stratified sea. Place in this sea a float that is neutrally buoyant at some depth. Slightly displace the float from this depth and let go. The float will then begin to oscillate about its equilibrium position. The frequency of oscillation is the local value of the buoyancy frequency.

As far as modelling is concerned, there are usually two choices. One can either model the sea as a series of layers or one can model the sea as continuously stratified. Only in this latter formulation is a description in terms of the Brunt-Väisälä frequency appropriate, for in a layered sea, vertical derivatives can only be represented as infinite spikes (Dirac δ functions) and these lack square roots. This is not to say that modelling the sea as comprising two layers is wrong, far from it. Merely that one should avoid modelling in terms of the buoyancy frequency. If a continuously stratified model is chosen, then commonly the density is assumed to take the form

$$\rho(x,y,z,t) = \rho_0(z) + \rho'(x,y,z,t) \qquad \text{where} \quad |\rho'| << |\rho_0|.$$

That the fluid is incompressible implies that

$$\frac{D\rho}{Dt} = \frac{\partial\rho}{\partial t} + u\frac{\partial\rho}{\partial x} + v\frac{\partial\rho}{\partial y} + w\frac{\partial\rho}{\partial z} = 0$$

which, given the small magnitude of horizontal gradients of density compared to vertical ones (except near fronts) reduces to

$$\frac{\partial \rho'}{\partial t} + w \frac{\partial \rho_0}{\partial z} = 0.$$

In the vertical, the balance is primarily hydrostatic, but when dealing with buoyancy effects, the non-hydrostatic correction is significant. The small deviation of pressure from hydrostatic is labelled p' and the vertical momentum equation is taken as largely hydrostatic, with the deviation from this obeying the balance

$$\rho_0 \frac{\partial w}{\partial t} = -\frac{\partial p'}{\partial z} - \rho' g.$$

It is this last term $(\rho' g)$ that is the restoring force that produces the oscillation once ρ' is eliminated. The result is

$$\frac{\partial^2 w}{\partial t^2} + N^2 w = -\frac{1}{\rho_0} \frac{\partial^2 p'}{\partial z \partial t}$$

from which the statement about the buoyancy frequency N is confirmed. The effects of Coriolis acceleration are not important here because the horizontal length scales are usually small. Where they are not, a more complicated formulation in terms of potential vorticity (see section 2.5) is necessary. Measurements in the ocean indicate that the density never departs by more than 2% from its mean value. Writing down the horizontal balance in its simplest form:

$$\rho_0 \frac{\partial u}{\partial t} = -\frac{\partial p'}{\partial x}$$

$$\rho_0 \frac{\partial v}{\partial t} = -\frac{\partial p'}{\partial y}$$

$$\frac{\partial u}{\partial x} + \frac{\partial v}{\partial y} + \frac{\partial w}{\partial z} = 0$$

and taking into account the smallness of the density variation gives the equation for w as

$$\frac{\partial^2}{\partial t^2} \left(\frac{\partial^2}{\partial x^2} + \frac{\partial^2}{\partial y^2} + \frac{\partial^2}{\partial z^2} \right) w + N^2 \left(\frac{\partial^2}{\partial x^2} + \frac{\partial^2}{\partial y^2} \right) w = 0.$$

See Gill (1982). If $N =$ constant a large variety of plane wave motions are possible.

It is the wind that blows across the surface of the sea that drives the currents in the upper layers. At some point, the buoyancy of the sea prevents the wind driven current from penetrating any further. The dimensionless ratio that alerts us to this is called the Richardson number, R_i and is defined as the ratio of the square of buoyancy frequency to shear:

$$R_i = \frac{N^2}{|\partial u / \partial z|^2} = -\frac{1}{\rho_0} \frac{\partial \rho_0}{\partial z} \Big/ \left| \frac{\partial U}{\partial z} \right|^2$$

where U is the current in the easterly (x) direction. The square avoiding troublesome square roots of density gradients. The above definition is often referred to as the local or gradient Richardson number in order to distinguish it from others. It is worth a little foray into stability theory here in order to show the origins of Richardson number criteria. This criteria will be stated later, so skip the mathematical detail if you wish. If current U is included in the mathematical description of the flow, then still neglecting rotational effects the equations are now

$$\rho_0 \left[\frac{\partial u}{\partial t} + U \frac{\partial u}{\partial x} + w \frac{dU}{dz} \right] = -\frac{\partial p'}{\partial x}$$

$$\rho_0 \left[\frac{\partial w}{\partial t} + U \frac{\partial w}{\partial x} \right] = -\frac{\partial p'}{\partial z} - \rho' g$$

$$\rho_0 \left[\frac{\partial \rho'}{\partial t} + U \frac{\partial \rho'}{\partial x} + w \frac{d\rho_0}{dz} \right] = 0$$

$$\frac{\partial u}{\partial x} + \frac{\partial w}{\partial z} = 0$$

where only motion in the $x-z$ plane is considered for simplicity. The last (continuity) equation leads immediately to a description in terms of a streamfunction:

$$u = -\frac{\partial \psi}{\partial z}, \quad \text{and} \quad w = \frac{\partial \psi}{\partial x}.$$

Elimination of variables is now possible in favour of the streamfunction ψ which after assuming the Boussinesq approximation (that the effects of density do not affect the horizontal momentum) gives the single equation

$$\left(\frac{\partial}{\partial t} + U \frac{\partial}{\partial x} \right)^2 \left(\frac{\partial^2}{\partial x^2} + \frac{\partial^2}{\partial z^2} \right) \psi + N^2 \frac{\partial^2 \psi}{\partial x^2} - \frac{d^2 U}{dz^2} \left(\frac{\partial}{\partial t} + U \frac{\partial}{\partial x} \right) \frac{\partial \psi}{\partial x} = 0.$$

At this juncture, we forward reference Chapter 6 and look for wave-like solutions with wave speed (celerity) c, and provided the wave speed is never the same as the current U the amplitude of the wave A where

$$\psi = A(z) e^{ik(x - ct)}$$

satisfies the differential equation

$$\frac{d^2 A}{dz^2} + \left[\frac{N^2}{(U(z) - c)^2} - \frac{d^2 U/dz^2}{U(z) - c} - k^2 \right] A = 0.$$

The key here is that the complex representation of the wave allows for growth and decay as well as oscillations. If this last differential equation is examined for stability, the term

$$\frac{1}{4} \left(\frac{dU}{dz} \right)^2 - N^2$$

occurs when the equation is integrated between two arbitrary levels. This leads to the following important conclusions:

1. If $R_i > 1/4$ the flow is stable to all small disturbances.

2. If $R_i < 1/4$ the flow may be stable or unstable, and in many cases it is unstable.

3. If $R_i > 1$ then buoyancy effects dominate and overall stability is assured (this can be seen from the definition of the Richardson number).

The more practical among you will be pleased to know that the above criteria are confirmed by experiment. The $1/4$ criteria is due to wave instability and observations confirm that these exist and can grow when the gradient Richardson number exceeds this value. The larger value $R_i = 1$ indicates where buoyancy suppresses shear, and observations indicate that instabilities (either waves or turbulent eddies) are suppressed once more. The Richardson number is closely linked to another dimensionless ratio called the Froude number (see Chapter 6) which in the context of this chapter is defined as the ratio U/c. The Froude number is important when considering waves and currents together in a river or estuary and hence features in civil engineering hydraulics models. The stability of waves in a current in a stratified fluid is governed by the magnitude of internal Froude numbers (where c is an internal wave speed), and this must be allied to stability of shear layers in a similarly stratified flow. The book by Baines (1995) takes this further and should be consulted by the interested reader. The text by Lewis (1997) tells more about the effects of stratification in the presence of turbulence as applied to estuaries should be consulted for further information.

2.4.3 Eddy Viscosity Models

In the previous two subsections dimensional analysis could not really be used effectively. Now however we certainly will use it but before doing so we have to propose a reasonable frictional law. The "friction" being discussed here is not molecular friction of the type met in mechanics when objects slide or do not slide down planes, but turbulence. Once again there is a choice for the reader to skip the details and accept the model or to follow the derivation in part or in whole. As there are learned books devoted to modelling fluid turbulence (e.g. the classical text by A A Townsend (1954)) only a brief introduction is given here. The basic idea in applying turbulence ideas to oceanography is that any current in the sea can be thought of as a mean flow that may be time dependent but is steady on a short time scale of a few seconds, plus a small random deviation. Mathematically $\mathbf{u} = \bar{\mathbf{u}} + \mathbf{u}'$ where $\overline{\mathbf{u}'} = 0$ where overbar denotes the mean taken over a few seconds. This is now inserted in all the governing equations and the overall time average (again over a few seconds) taken. All is as expected, except for the non-linear advective acceleration which produces the extra non zero term

$$\overline{(\mathbf{u}'.\nabla)\mathbf{u}'}$$

on the left hand side of the Navier Stokes equations. This small random deviation is of course unknown, but fortunately the non linear non zero term is the only place in the equations where it occurs. It is called the Reynolds stress term and although it may look harmless, it actually contains 9 terms consisting of all pairs of combinations of the components of $\mathbf{u}' = (u', v', w')$ with itself. One of the simplest assumptions to make is to relate appropriate parts of this Reynolds stress to gradients in the mean flow. The Reynolds stress is in fact a tensor (strictly a second order tensor) which is made up from the nine components and can be represented by a 3×3 matrix. It can be succinctly written τ_{ij} where the suffices i and j independently run from 1 to 3. In detail

$$
\begin{pmatrix}
\tau_{11} & \tau_{12} & \tau_{13} \\
\tau_{21} & \tau_{22} & \tau_{23} \\
\tau_{31} & \tau_{32} & \tau_{33}
\end{pmatrix}
=
\begin{pmatrix}
-\rho \overline{u'^2} & -\rho \overline{u'v'} & -\rho \overline{u'w'} \\
-\rho \overline{v'u'} & -\rho \overline{v'^2} & -\rho \overline{v'w'} \\
-\rho \overline{w'u'} & -\rho \overline{w'v'} & -\rho \overline{w'^2}
\end{pmatrix}
$$

where each τ_{ij} in fact measures the covariance (statistical measure of agreement) between two of the fluctuating components $\mathbf{u}' = (u', v', w')$. If $i = j$ it is the autocovariance of u', v' or w' that is being measured. Both matrices are of course symmetric ($\tau_{ij} = \tau_{ji}$). This follows straight away from the right hand matrix. In fact if $i = j$ then the stress is called a normal stress. Since normal stresses act in a similar fashion to pressure, they can be safely overlooked (or more accurately absorbed by the pressure). When $i \neq j$ the stress is a shear stress and it is these that need to be modelled. It is of course the *gradients* of the stresses that contribute to the conservation of momentum in much the same way as does the gradient of pressure and not the pressure itself. Turbulence and friction have in common the transfer of momentum in a direction at right angles to the flow. The magnitude of the rate at which this transfer takes place in a turbulent ocean current is what is attempted to be modelled by eddy viscosity. The big assumptions are one that such a relationship is reasonable and two that the relationship is linear. Neither of these can be justified except perhaps *a posteriori*. To home in on a particular component, let us choose $\tau_{13} = \tau_{31}$. In this case, the eddy viscosity assumption, a turbulence equivalent to assuming a Newtonian viscous fluid, leads to

$$
\tau_{31} = -\rho \overline{w'u'} = -\rho \nu_v \frac{\partial \bar{u}}{\partial z},
$$

where ν_v is the eddy viscosity or Austauch coefficient. It is the turbulent equivalent to kinematic (not dynamic) viscosity. Not only is it much bigger than normal viscosity, usually by a factor of 10^6 but it is of course not a fixed property of the fluid. The gradient of the stress and not the stress itself gives the net force. No apology is necessary for reiterating that this is the same as gradients of pressure not pressure itself causing net force. Hence if the overall force balance is to include shear stress effects then terms such as

$$
\frac{\partial}{\partial z} \tau_{31} = \frac{\partial}{\partial z} \left(\nu_v \frac{\partial u}{\partial z} \right)
$$

need to be included. If the eddy viscosity ν_v is a constant, then this can be written in terms of a second derivative:

$$
\nu_v \frac{\partial^2 u}{\partial z^2}.
$$

The stress τ_{31} is a stress that represents the shear due to a current travelling in an easterly direction in the presence of either the sea bed or the sea surface. Other shear stresses, for example τ_{21} would represent the shear due to an easterly current near a north south coast, could also be put in terms of gradients in mean flow:

$$\tau_{21} = -\rho \overline{v'u'} = -\rho \nu_H \frac{\partial \bar{u}}{\partial y},$$

but this time the eddy viscosity ν_H is representative of the *horizontal* transfer of momentum in the current. The horizontal eddies that effect this transfer are correspondingly larger than the vertical counterparts, hence $\nu_H \gg \nu_v$ (by a factor of about 10^4 in fact). Let us examine the equation of motion (conservation of momentum) in the x-direction. If we include advection, vertical and horizontal momentum transfer (eddy viscosity) then dimensional analysis can be performed on this equation. As all reference to primed (fluctuating) quantities has ceased, the overbar can be dropped. The general x wise equation of motion then is rather daunting, until one remembers that is not actually going to be solved. Here it is:

$$\frac{\partial u}{\partial t} + u\frac{\partial u}{\partial x} + v\frac{\partial u}{\partial y} + w\frac{\partial u}{\partial z} - fv = -\frac{1}{\rho}\frac{\partial p}{\partial x} + \nu_v \frac{\partial^2 u}{\partial z^2} + \nu_H \left(\frac{\partial^2 u}{\partial x^2} + \frac{\partial^2 u}{\partial y^2}\right).$$

This is amenable to dimensional analysis of the type that estimates magnitudes. This is the second role of dimensional analysis mentioned earlier. The first step is to non-dimensionalise all variables as follows; write

$$u = Uu', \qquad v = Uv', \qquad w = Ww' = U\frac{D}{L}w',$$

$$(x,y) = L(x',y'), \qquad z = Dz', \qquad t = \frac{t'}{f_0}, \qquad f = f_0 f', \qquad p = p' f_0 UL.$$

Here capital letters carry the magnitude and are constant, whilst primed quantities are variable but their magnitude is around unity. As messy as it looks, this is not in fact the most general non dimensionalisation as the following assumptions have already been made. First of all, the pressure and Coriolis terms are assumed to be the same order of magnitude which is equivalent to assuming that the underlying balance is geostrophic. Secondly, the appropriate time scale is assumed to be about a pendulum day $1/f_0$, which means the motion is what is termed quasi-geostrophic or on the time scale of about a day or possibly larger. Finally, the continuity equation has been used to determine the magnitude of the vertical motion W. It is certainly general enough for us here. The primed variables are substituted into the equation of motion, and the result is divided by $f_0 U$ which gives

$$\frac{\partial u}{\partial t} + \frac{U}{f_0 L}\left(u\frac{\partial u}{\partial x} + v\frac{\partial u}{\partial y} + w\frac{\partial u}{\partial z}\right) - fv = -\frac{1}{\rho}\frac{\partial p}{\partial x} + \frac{\nu_v}{f_0 D^2}\frac{\partial^2 u}{\partial z^2} + \frac{\nu_H}{f_0 L^2}\left(\frac{\partial^2 u}{\partial x^2} + \frac{\partial^2 u}{\partial y^2}\right)$$

where all the dashes have been dropped for convenience. From here the reasoning goes as follows. Since the non dimensional quantities u, v, w, x, y, z, t and f

are all assumed to be near unity in magnitude, as are all gradients, the dimensionless combinations $\dfrac{U}{f_0 L}$, $\dfrac{\nu_v}{f_0 D^2}$ and $\dfrac{\nu_H}{f_0 L^2}$ emerge as indicating the relative importance of various terms compared to the Coriolis acceleration. That no dimensionless groupings appear in front of either the local acceleration or the pressure terms is a consequence of the assumptions. The combination $\dfrac{U}{f_0 L}$ is called the Rossby number, R_O after the pioneering Swedish then US (he moved to the USA in 1926) oceanographic and atmospheric modeller Carl-Gustaf Arvid Rossby (1898 - 1957). If R_O is about unity then the advective acceleration is similar in magnitude to the Coriolis acceleration and hence cannot be ignored in any model. The way to proceed is if it is suspected that R_O may be significant, representative values of the parameters U, f_0 and L are inserted into the formula for the Rossby number. For example, for general ocean circulation in mid-latitudes, $U = 0.1 ms^{-1}$, $f_0 = 10^{-4} s^{-1}$ and $L = 10^6 m$ which gives a Rossby number $R_O = 10^{-3}$ which is small enough to be ignored. Hence the advective terms can be overlooked when modelling the general circulation of the ocean. Of course this only remains true provided the above magnitudes remain representative. If variables change on a length scale of $10^4 m$ this becomes the new L, and if at the same time currents can be as large as $1 ms^{-1}$, then the Rossby number becomes unity and here the advective terms can no longer be ignored. This is what happens near the detached western boundary currents as they cross the northern parts of the ocean (Gulf Stream - Atlantic, Kuroshio - Pacific). Another circumstance where the Rossby number may be unity is strong tides around headlands; try putting in some typical magnitudes yourself. Of course, in both of these regimes, frictional (turbulence) effects may also be significant. The other two groups of dimensionless combinations $\dfrac{\nu_v}{f_0 D^2}$ and $\dfrac{\nu_H}{f_0 L^2}$ indicate the relative importance of vertical and horizontal turbulent transfer of momentum (friction) respectively. $\dfrac{\nu_v}{f_0 D^2} = E_v$ is the vertical Ekman number and $\dfrac{\nu_H}{f_0 L^2} = E_H$ the horizontal Ekman number. These are named after the Swedish physicist Vagn Walfrid Ekman (1874 - 1954) often wrongly attributed to be Norwegian; he worked in the University of Lund in Sweden where he was Professor of Mechanics and Mathematical Physics, he was the son of a Swedish physical oceanographer Frederick Laurentz Ekman in Stockholm and died at Gostad in Sweden and although he spent much of his working life surrounded by Norwegians such as Bjerknes and Nansen, he was definitely Swedish! It is Ekman who first formulated a simple surface wind driven current model, seeking to explain the observations of Nansen as his ship the Fram was icebound in his attempt to reach the North Pole (1893 - 1896). We shall explore this model later in this section. Indeed, you would be right to expect that vertical transfer of momentum assumes some importance near the sea surface and near the sea bed. Horizontal transfer of momentum might be expected to be important in regions where substantial currents rub against the coast, for example Florida (Gulf Stream) or East Africa (Somali Current). The difficulty in using dimensional analysis here is, unlike for the Rossby number, there are quantities ν_v and ν_H that are not directly measurable. Thus it is not easy to home in on an uncontroversial magnitude for either of the Ekman numbers. Nevertheless there are accepted values for eddy viscosity in some circumstances and the concept of the Ekman number is useful for modellers.

What we have seen here therefore is the use of dimensional analysis to tell us what the dominant processes might be in one equation. Once this has been done, the equation is simplified by the elimination of unwanted terms, but it still needs to be solved. The solution may be possible analytically, and the surface Ekman layer model is one of these. More likely, numerical methods will need to be employed, and this is the subject of the next chapter. The non-linear terms often provide a bar to the analytical solution to problems where the Rossby number has a magnitude of around unity. The concept of vorticity is an aid here as it provides another tool, another way of looking at what makes the ocean "tick". It is time to introduce this now.

2.4.4 Vorticity

Vorticity is a very important fundamental quantity, and has a special place in oceanography. Understanding about it is well worth the effort. It is the fluid counterpart of angular momentum, which is the quantity preserved by spinning ice skaters; as the ice skater pirouettes and draws in the arms, the rate of spin increases. The concentration of the mass of the skater nearer the centre of mass by drawing in the arms and the increase in angular speed of the skater, balance themselves out ensuring that the angular momentum remains constant. In mechanics, the angular momentum of a body is a quantity that tends not to vary. This fact lies behind the behaviour of the planets and satellites outside the earth and tops and gyroscopes closer to home. In fluid mechanics, the vorticity has an analogous role. We have not defined angular momentum for to do so would be too much of a digression into mechanics. We will however define vorticity, but not just yet suffice to that it is the spatial derivative of velocity, giving it the dimensions of T^{-1}. It's stark definition in terms of mathematically well defined quantities (the curl of the velocity vector) is singularly unilluminating. Instead, we discuss some properties of a fluid first. If a river or stream flows at a constant rate, then near to the bank there is generation of vorticity. If a neutrally buoyant float with a line drawn on its topmost surface was placed in the stream next to the bank, it would drift with the stream, but because the side nearest the bank was always travelling slower than the opposite side (due to the friction of the bank) the line would rotate. This rotation is a measure of vorticity. Paradoxically the same float put near the vortex of a draining bath would not rotate at all (until it hit the vortex itself); the vorticity of a fluid near a vortex is zero. Now the vorticity of a fluid tends to be preserved by motion not influenced by friction, and many properties of quite complex looking flows can be deduced from following the vorticity and assuming it is constant. In oceanography, an important source of vorticity arises from the change of the Coriolis parameter with latitude (recall that it is zero at the equator but $\pm 2\Omega$, where Ω is the angular speed of the earth at each pole). The vorticity arising from the Coriolis acceleration only depends on the position of a current on the earth's surface and so is distinctive. The total vorticity of a large scale ocean current has three influences: the local changes ("shear") in the current; the local value of the Coriolis parameter; and the ocean depth. The whole is given the distinctive name *planetary vorticity*, and its conservation is crucial to the full understanding of ocean scale physics. It is less crucial to us who are more concerned with smaller scale modelling appropriate to coasts, but it is still well worth discussing for the insight it gives. For those who are not mathematically inclined, only the result and applications of it need be of concern. For completeness however a derivation follows which is more technical than virtually the whole of the rest of the text.

In order to proceed with even the statement of the conservation of potential vorticity we need to define the quantity

$$\frac{D}{Dt}$$

which is the derivative following the fluid. This is the acceleration "following the fluid", that is from the point of view of a particle moving with the fluid. The combination of local acceleration and advective acceleration is

$$\frac{D\mathbf{u}}{Dt} = \frac{\partial \mathbf{u}}{\partial t} + (\mathbf{u}.\nabla)\mathbf{u},$$

so the derivative following the flow or *total derivative* of the flow itself is simply the two fluid accelerations (point or local, and advective) combined. In some respects, such a description is more natural than the use of fixed axes, but the penalty is very awkward expressions for quantities such as dissipation due to turbulence. Even the simplest model (constant eddy viscosity) involves embedded Jacobians! This is because in order to use Newton's laws, all has to be referred back to a start position. Nevertheless particle tracking descriptions are very useful for modelling diffusion and other systems where water borne material is being carried along with the flow. This description is given the name *Lagrangian* whereas the description relative to fixed axes is called *Eulerian*. We shall almost always use Eulerian descriptions. Using the total derivative notation, the two horizontal equations of motion become

$$\frac{Du}{Dt} - fv = -\frac{1}{\rho}\frac{\partial p}{\partial x}$$

$$\frac{Dv}{Dt} + fu = -\frac{1}{\rho}\frac{\partial p}{\partial y}$$

where all frictional effects are ignored. In order to eliminate the pressure, one takes the y derivative of the first equation and the x derivative of the second and subtracts. This is equivalent to "taking the curl" of the vector equation of motion, as the more mathematically minded of you will have spotted, but no matter. More important to realise is that because the total derivative is not linear, the cross differentiation has to be done carefully term by term. The details of the algebra are omitted, but the result is

$$\frac{\partial}{\partial x}\frac{Dv}{Dt} - \frac{\partial}{\partial y}\frac{Du}{Dt} = \frac{D}{Dt}\left(\frac{\partial v}{\partial x} - \frac{\partial u}{\partial y}\right) + \left(\frac{\partial v}{\partial x} - \frac{\partial u}{\partial y}\right)\left(\frac{\partial u}{\partial x} + \frac{\partial v}{\partial y}\right).$$

The quantity

$$\zeta = \left(\frac{\partial v}{\partial x} - \frac{\partial u}{\partial y}\right)$$

is the vorticity of the (two dimensional) flow. The result of the cross differentiation is therefore

$$\frac{D\zeta}{Dt} + \zeta\left(\frac{\partial u}{\partial x} + \frac{\partial v}{\partial y}\right) + \frac{\partial fu}{\partial x} + \frac{\partial fv}{\partial y} = 0$$

which can be written

$$\frac{D}{Dt}(\zeta + f) + (\zeta + f)\left(\frac{\partial u}{\partial x} + \frac{\partial v}{\partial y}\right) = 0.$$

(This uses the fact that

$$\frac{Df}{Dt} = v\frac{\partial f}{\partial y}$$

which of course is actually zero for models with constant f.) The quantity $\zeta + f$ is referred to as the absolute vorticity. In order to eliminate the horizontal divergence term $\left(\frac{\partial u}{\partial x} + \frac{\partial v}{\partial y}\right)$ it is necessary to do some manipulating with the equation of continuity, namely vertically integrating it.

The continuity equation in three dimensions is

$$\frac{\partial u}{\partial x} + \frac{\partial v}{\partial y} + \frac{\partial w}{\partial z} = 0.$$

Now if it is assumed that the sea surface is given by the equation $z = \eta(x, y, t)$ and the sea bed by the equation $z = -h(x, y)$, then integrating vertically between these two surfaces assuming that neither u nor v depend on z gives

$$(\eta + h)\left(\frac{\partial u}{\partial x} + \frac{\partial v}{\partial y}\right) + \left[w\right]_{-h(x,y)}^{\eta(x,y,t)} = 0.$$

Realising that at the surface $w = \dfrac{D\eta}{Dt}$ due to fluid particles at the surface always remaining there (kinematic condition) and that at the bed $w = -\dfrac{Dh}{Dt}$ (no flow through the bed itself), and writing $H(x, y, t) = \eta(x, y, t) + h(x, y)$ to represent the total height gives the alternative continuity equation

$$\frac{\partial u}{\partial x} + \frac{\partial v}{\partial y} = -\frac{1}{H}\frac{DH}{Dt}.$$

Inserting this into the above vorticity equation gives the simple (looking) equation

$$\frac{D}{Dt}\left(\frac{\zeta+f}{H}\right) = 0.$$

This is the standard form of the potential vorticity equation. If you did not like the derivation (who does) or more importantly did not understand it, then take it as true and let us see some consequences that arise from it. In those parts of the ocean where there is little fluid shear, the local value of ζ is virtually zero. At these places, the vertically integrated flow follows contours of f/H (called potential vorticity contours). In the early days of oceanography either side of the second world war, this fact was used extensively to infer deep circulation. In particular that there must be a deep return flow under the Gulf Stream from Cape Hatteras to the Florida Keys along the continental slope.

We now have enough modelling tools to execute model studies. These case studies form the final section of this chapter. First however let us take a brief look at time dependent effects, although this is necessarily brief because most time dependence is manifest via waves which are not met formally until chapter 6.

2.5   Unsteady Ocean Dynamics

A little thought will tell you that potential vorticity balance ought to be useful in studying the unsteady dynamics of the ocean, and so it proves. The simplest (unforced) equations are purely two dimensional and hydrostatic and are governed by the equations

$$\frac{\partial u}{\partial t} - fv = -g\frac{\partial \eta}{\partial x}$$

$$\frac{\partial v}{\partial t} + fu = -g\frac{\partial \eta}{\partial y}$$

$$\frac{\partial \eta}{\partial t} + h\left(\frac{\partial u}{\partial x} + \frac{\partial v}{\partial y}\right) = 0,$$

which will be called the Laplace Tidal Equations in Chapter 6 but are in fact a simplification using local cartesian co-ordinates due to Lord Kelvin (William Thompson (1824 - 1907)), published in 1869. Laplace's original 1776 version used spherical polar co-ordinates! An equation for η can be derived which contains the potential vorticity ζ, namely

$$\frac{\partial^2 \eta}{\partial t^2} - gh\left(\frac{\partial^2 \eta}{\partial x^2} + \frac{\partial^2 \eta}{\partial y^2}\right) + f^2 h\zeta = 0.$$

This is essentially a wave equation. Waves belong to Chapter 6 and are introduced properly there. Nothing is done with this unsteady model here.

There are layered models that describe the general motion of a wind driven stratified sea. These are usually large scale and so are not dwelt on in detail here, but the methodology is interesting. If it is assumed that the ocean is still governed by linear quasi-geostrophic dynamics but this time forced by the wind stress, the horizontal equations retaining pressure explicitly are

$$\frac{\partial u}{\partial t} - fv = -\frac{1}{\rho_0}\frac{\partial p}{\partial x} + \frac{1}{\rho_0}\frac{\partial \tau^x}{\partial z}$$

$$\frac{\partial v}{\partial t} + fu = -\frac{1}{\rho_0}\frac{\partial p}{\partial y} + \frac{1}{\rho_0}\frac{\partial \tau^y}{\partial z}$$

where ρ_0 is a reference density (called ρ earlier when density was constant) the other symbols retain earlier meanings; (τ^x, τ^y) is the wind stress. Gill and Clarke (1974) used an elegant method employing expansion of variables in normal modes as follows:

$$p = \sum_{n=0}^{\infty} p'_n(x,y,t)\hat{p}_n(z), \quad \text{and} \quad \rho_0 g(u,v) = \sum_{n=0}^{\infty} (u'_n(x,y,t), v'_n(x,y,t))\hat{p}_n(z).$$

This form when substituted into the quasi-geostrophic equations neatly reproduces quasi-geostrophic equations for each mode as follows

$$\frac{\partial \hat{u}_n}{\partial t} - f\hat{v}_n = -g\frac{\partial \hat{p}_n}{\partial x} + \hat{X}_n$$

$$\frac{\partial \hat{v}_n}{\partial t} + f\hat{u}_n = -g\frac{\partial \hat{p}_n}{\partial y} + \hat{Y}_n$$

where we have also decomposed the wind stress into the same modes via

$$g\left(\frac{\partial \tau^x}{\partial z}, \frac{\partial \tau^y}{\partial z}\right) = \sum_{n=0}^{\infty} (\hat{X}_n, \hat{Y}_n)\hat{p}_n(z).$$

The continuity equation is not written down and is in general very complex. The analysis of such model as this again involves waves so this is not done here. In Chapter 6, similar types of modelling strategies are adopted, however the problem with the above model is that there is no easy way to incorporate horizontal boundaries.

It is pertinent to ask whether there are potential vorticity arguments which can be used when the flow is baroclinic. The answer is yes, but with difficulty. The vorticity vector ζ is defined by

$$\zeta = \nabla \times \mathbf{u} = \left(\frac{\partial v}{\partial z} - \frac{\partial w}{\partial x}, \frac{\partial v}{\partial z} - \frac{\partial w}{\partial y}, \frac{\partial v}{\partial x} - \frac{\partial u}{\partial y}\right)$$

and would no longer be solely in the the z direction. Detailed treatment is not given here (see Gill (1982) for a clear but necessarily quite mathematical account), but here are a few snippets of information. Vorticity in a stratified fluid is generated by the presence of angles between pressure surfaces and density surfaces. The non vanishing of this angle is in fact a rigorous definition of the term *baroclinic*. The baroclinicity vector $\mathbf{B}$ is defined by

$$\mathbf{B} = \frac{1}{\rho^2}\nabla\rho \times \nabla p$$

and vorticity will be generated where for example there are horizontal density gradients (given that pressure will be close to hydrostatic, hence near horizontal pressure surfaces). We have already seen that horizontal density gradients induce changes in velocity with depth; the thermal wind equations, but this mechanism depends on rotation via the Coriolis parameter whereas the vorticity generated by the baroclinicity vector is independent of rotation. In fact, baroclinicity induced vorticity is often key in small scale flows, including the study of stratified flows in the laboratory as well as estuaries and small seas where the Coriolis effect can be ignored.

2.6 Case Studies

2.6.1 Ocean Surface Layer Models

The first models of surface layer physics consisted of a simple balance between Coriolis acceleration and vertical transfer of momentum. From the dimensional considerations of section 2.4.3 this implies that the vertical Ekman number

$$E_v = \frac{\nu_v}{fD^2}$$

must be of order 1. The values $f = 10^{-4}s^{-1}$, $D = 10^3 m$ and $\nu_v = 10^2 m^2 s^{-1}$ are appropriate, which does indeed give $E_v = 1$. When actually solving equations as distinct from using them for dimensional analysis one has the choice of either working with dimensional or dimensionless variables. Both have their merits, but in the present context the use of dimensional variables is less confusing. The equations for steady Ekman flow, called the Ekman equations are

$$-fv = \nu_v \frac{\partial^2 u}{\partial z^2}$$
$$fu = \nu_v \frac{\partial^2 v}{\partial z^2}.$$

Before presenting the solution, one should ask what has happened to the pressure terms. If one writes the total current as $(u + u_g, v + v_g)$ then we have

$$-f(v + v_g) = -\frac{1}{\rho}\frac{\partial p}{\partial x} + \nu_v \frac{\partial^2}{\partial z^2}(u + u_g)$$
$$f(u + u_g) = -\frac{1}{\rho}\frac{\partial p}{\partial y} + \nu_v \frac{\partial^2}{\partial z^2}(v + v_g)$$

and we set (u_g, v_g) to balance the pressure gradient. Therefore since (u_g, v_g) does not depend on z, (u, v) satisfy the Ekman equations as desired.

As the Ekman equations are differential equations, boundary conditions are required for a complete solution. Let us look at what these might be. Ekman flow is driven by the wind, therefore far beneath the sea surface (u, v) will tend to zero. At the sea surface one makes the assumption that the stress is the same as the wind stress. So if the wind stress is given by the horizontal vector (τ^x, τ^y) then

$$\tau_{31} = \tau^x, \qquad \tau_{32} = \tau^y.$$

Hence, at the sea surface

$$(\tau^x, \tau^y) = -\rho\nu_v \left(\frac{\partial u}{\partial z}, \frac{\partial v}{\partial z} \right).$$

With the other assumptions already made (neglecting the non linear advection terms arising from assuming small Rossby number) the sea surface can be taken as $z = 0$. This is sometimes called the "rigid lid" approximation and amounts to the suppression of surface waves. Any sea surface slope due to geostrophic effects is included in (u_g, v_g) and may be added later. The Ekman equations plus boundary conditions form a boundary value problem with a unique solution. In particular a unique solution which can be written in closed analytical form as follows

$$(u, v) = V_0 e^{\pi z/D_0} \left(\pm\cos\left(\frac{\pi}{4} + \frac{\pi}{D_0}z \right), \sin\left(\frac{\pi}{4} + \frac{\pi}{D_0}z \right) \right),$$

where the constants V_0 and D_0 have been introduced for convenience and are defined as follows:

$$V_0 = \frac{\pi}{\rho D_0 |f|} \sqrt{2(\tau^x)^2 + 2(\tau^y)^2}$$
$$D_0 = \pi\sqrt{2\nu_v/|f|}$$

$|f|$ is the absolute value (magnitude) of f which accounts for whether we are north or south of the equator. The plus sign is taken north of the equator, the minus sign south of it. The constant V_0 is termed the "total Ekman surface current" and the constant D_0 is called the "depth of frictional influence". Both terms are reasonably self explanatory. Here we have more or less followed Pond and Pickard "Introductory Dynamical Oceanography" (1983) Pergamon Press. There are equally valid solutions that differ in detail (positions of $\sqrt{2}$ or π) but the essentials are the same. Now that we have a solution, we can validate it via various observations.

The classical reason for the development of the Ekman layer model was in response to observations made by the Norwegian biological oceanographer, Arctic explorer, statesman, humanitarian and now national hero, Fridtjof Nansen (1861 - 1930) whilst he was icebound during a three year (1893 - 1896) voyage. It was a heroic attempt to be the first to reach the North Pole. His ship (the Fram) was specifically designed to withstand the pressures of the ice as it froze, in fact the freezing water lifted the Fram so that the ship rode on its frozen surface. He got close, but did not succeed in reaching the pole. The Fram is now preserved in a museum across the water from the city of Oslo with easy access by ferry and is well worth a visit. One observation that surprised Nansen was that the surface ice movement did not follow the driving wind but seemed to move to the right of it. Nansen suggested between 20^0 and 40^0. The Ekman model gives 45^0. It also gives a net flow 90^0 to the right (left) of the wind in the northern (southern) hemisphere. This feature can be used to explain why a flow along a coast can give rise to vertical currents at the coast (upwelling) which is important for biological productivity. For a west facing coast, a wind along the coast from the south in the southern hemisphere or from the north in the northern hemisphere will induce the surface flow to be offshore according to the Ekman model. In order to replace this surface water that moves out towards the sea, water which is usually colder has to rise from the depths. This is the upwelling and this cold water, being nutrient rich provides food for local plankton which in turn feed the fish. This model of upwelling first saw the light of day in 1908 and although there are considerably better more complex models, the underlying reason behind upwelling in places like Nigeria and Peru is still very much the Ekman picture. The depth of frictional influence (D_0 above) is of the order of 10 or 20 metres in the mid latitudes, so Ekman flow (if it is a valid model) will describe flow very near and at the sea surface. Ekman spirals are seldom observed, in fact they were discounted as just an ideal picture, almost fictional, for years until the more sensitive non invasive instruments (laser doppler anemometry) of the 1970s could pick them out. In general the surface waves and unsteady nature of the wind will destroy Ekman layer structure. A three dimensional diagram of the solution to the Ekman equations in the northern hemisphere will reveal the current to be a spiral, in a direction 45^0 to the right of the wind at the sea surface and spiralling away from the direction of the wind the deeper one goes. All the while, the amplitude of the current diminishing until at a depth corresponding to D_0 the current has all but vanished. The solution is often called the Ekman spiral for this reason. South of the equator $f < 0$ and so the spiral starts off 45^0 to the *left* of the wind at the sea surface and spirals away. One simple alternative is to propose a depth dependent eddy viscosity. This could be justified as representing the change in turbulence characteristics the further one gets from the sea surface. One simple model using a quadratic dependence was used by Dyke (1977) and predicted a smaller angle of deviation for the surface stress (10°). The total depth integrated current is always 90° for eddy viscosity models, no matter how the eddy viscosity depends on depth. One should ask about different models of wind driven currents. There are many. Oceanographers recognise the existence of a well mixed surface layer virtually always present in cooler climes and virtually always due to the action of the wind on the sea surface, and it is tempting to equate this with the Ekman layer. However to do so would invariably be wrong as the base of the mixed layer called the thermocline (more correctly the pycnocline) is very stratified. It is an interface between well mixed surface water and less well mixed deeper water and since Ekman layer physics ignores all density changes it could not possibly say anything about where this interface might be. Commonly, if the Ekman layer occupies the top 20 metres of the sea, then the well mixed layer extends to about 100 or 200 metres. Models of the well mixed layer will include unsteady effects (Ekman's original model included the time dependent terms) more sophisticated turbulence models as well as thermal forcing and possibly advective acceleration.

2.6.2 Analysing Western Boundary Currents

The Ekman layer at the surface of the sea has just been discussed under the assumption of constant Coriolis parameter. If we now assume a β plane approximation whereby

$$f = f_0 + \beta y \quad \text{and} \quad \frac{\partial f}{\partial y} = \beta$$

a useful result can be derived. Starting with the geostrophic equations:

$$-fv = -\frac{1}{\rho}\frac{\partial p}{\partial x}$$

$$fu = -\frac{1}{\rho}\frac{\partial p}{\partial y},$$

differentiate the first with respect to y, the second with respect to x and subtract to obtain

$$\frac{\partial}{\partial y}(fv) + \frac{\partial}{\partial x}(fu) = 0$$

or, assuming the β plane approximation

$$\beta v + f\left(\frac{\partial u}{\partial x} + \frac{\partial v}{\partial y}\right) = 0.$$

The equation of continuity is

$$\frac{\partial u}{\partial x} + \frac{\partial v}{\partial y} + \frac{\partial w}{\partial z} = 0$$

from which the horizontal divergence terms can be eliminated to give

$$\beta v = f\frac{\partial w}{\partial z}.$$

This expresses the conservation of potential vorticity in areas free of friction (turbulent momentum transfer) and is called *Sverdrup balance*. Vertically integrate the Sverdrup balance from the bottom of the sea to just outside the bottom of the sea surface Ekman layer, and we deduce that

$$\int_{\text{sea bed}}^{\text{Ekman layer}} v\,dz = \frac{f}{\beta}W_E,$$

where W_E is the vertical velocity at the outer edge of the Ekman layer. Due to divergence within the Ekman layer itself, and the character of the wind driving at the sea surface, W_E is in a downwards direction. This implies that the net flow in the entire interior of the ocean is southerly. The demands of continuity mean that there has to be a northerly flow, and this flow is concentrated in places where turbulent momentum transfer cannot be neglected. Such a place is the western edge of the ocean, and it a model of this that now attracts our attention. It is now pertinent to ask what is the appropriate model of a coastal current. Observations confirm the above deduction that these coastal currents are on western sides of oceans (Gulf Stream, Kuroshio, Somali Current etc.). This is not immediately obvious, but arises from considering conservation of planetary vorticity and is confirmed by the model described later. Using an eddy viscosity description of turbulence means that we can write

$$\tau_{21} = -\rho\nu_H\overline{u'v'} = -\rho\nu_H\frac{\partial u}{\partial y}$$

$$\tau_{11} = -\rho\nu_H\overline{u'^2} = -\rho\nu_H\frac{\partial u}{\partial x}$$

where ν_H is assumed to be a constant eddy viscosity. For simplicity we also assume that the sea surface is flat so that the flow is non-divergent. This might seem at odds with Sverdrup balance which needs a divergence, but not so. The v that arises from $\partial w/\partial z$ is very small in this coastal region primarily because of the restricted fetch and can be readily neglected, hence we take

$$\frac{\partial u}{\partial x} + \frac{\partial v}{\partial y} = 0.$$

This enables us to substitute a single variable, the streamfunction ψ, instead of two variables u and v through the equations

$$u = \frac{\partial \psi}{\partial y}, \qquad v = -\frac{\partial \psi}{\partial x}.$$

We shall use this later. The horizontal equations of motion are a balance between Coriolis acceleration, pressure gradient and horizontal Reynolds stress, the latter being put in terms of gradients of the mean flow through the above eddy viscosity assumptions. This is similar to Ekman dynamics, the crucial difference being that the Reynolds stress terms now involve horizontal gradients rather than vertical gradients. This means that direct elimination of the pressure by assuming the velocity can be written as $(u + u_g, v + v_g)$ is not possible this time. Instead, we cross differentiate as when we derived the vorticity. The equations are

$$-fv = -\frac{1}{\rho}\frac{\partial p}{\partial x} + \nu_H\left(\frac{\partial^2 u}{\partial x^2} + \frac{\partial^2 u}{\partial y^2}\right)$$

$$fu = -\frac{1}{\rho}\frac{\partial p}{\partial y} + \nu_H\left(\frac{\partial^2 v}{\partial x^2} + \frac{\partial^2 v}{\partial y^2}\right)$$

and cross differentiation (first equation with respect to y, second equation with respect to x and subtract) eliminates the pressure, but at the expense of introducing higher derivatives. Of course we get the vorticity equation for this model which is

$$v\frac{\partial f}{\partial y} = \nu_H\left(\frac{\partial^2}{\partial x^2} + \frac{\partial^2}{\partial y^2}\right)\left(\frac{\partial u}{\partial y} - \frac{\partial v}{\partial x}\right).$$

Now, in mid-latitudes the β-plane approximation previously introduced can be invoked. With prescient knowledge of the streamfunction, this leads to the vorticity equation for the single variable ψ:

$$-\beta \frac{\partial \psi}{\partial x} = \nu_H \nabla^4 \psi,$$

where

$$\nabla^4 \psi = \frac{\partial^4 \psi}{\partial x^4} + \frac{\partial^4 \psi}{\partial x^2 \partial y^2} + \frac{\partial^4 \psi}{\partial y^4}.$$

It is this vorticity equation that is used to deduce that only western coasts can possess the intense boundary current, but we will not do this here. Instead, let us use this equation to deduce something about the dimensions. Close to the coast, variations *along* the coast will be much less than variations *perpendicular* to the coast, therefore it is appropriate to use the approximation

$$\frac{\partial}{\partial x} >> \frac{\partial}{\partial y}$$

which in turn means that, in this region

$$\beta \frac{\partial \psi}{\partial X} = \nu_H \frac{\partial^4 \psi}{\partial X^4},$$

where we have written $X = -x$ for convenience. Analytical solutions of this were first developed in 1950 by the US oceanographer Walter H Munk (1917 - 1999). Originally Austrian, Walter Munk moved to the States and eventually became the Director of the Scripps Institute for Oceanography having worked there all his life. It is the difference between having a minus sign and having no minus sign on the left that is responsible for the asymmetry (boundary current in the west, no boundary current in the east), but here we shall use this equation for dimensional purposes only. If L is the appropriate X length scale, then in dimensional terms,

$$\frac{\beta}{L} \sim \frac{\nu_H}{L^4}, \quad \text{or} \quad L \sim \left(\frac{\nu_H}{\beta} \right)^{1/3}.$$

Let us now substitute typical magnitudes. Munk's values are: $\beta = 1.9 \times 10^{-11} m^{-1} s^{-1}$, $\nu_H = 5 \times 10^3 m^2 s^{-1}$ which gives $L = (5 \times 10^3 / 1.9 \times 10^{-11})^{(1/3)} = 6 \times 10^4 m = 60 km$.

It is reasonable to ask whether there are other workable models of western boundary currents. Indeed there are. The one due to Henry Stommel (1920 - 1992) the much respected US oceanographer and meteorologist, is based not on lateral friction, but on the non linear advective acceleration balancing Coriolis acceleration. It can thus be deduced directly from the conservation of potential vorticity which holds for models that do not involve friction. In Stommel's model, the intense northwards flow of the Gulf Stream v is geostrophic and balanced by a sea surface slope. However, in order to be realistic, the ocean is assumed to be two layered with the Gulf Stream confined to the upper layer.

In two layer models, the density difference between the layers is small, which means that the depth of the upper layer has to vary a lot in order for geostrophy to hold. See later for actual magnitudes. Let the density of the top layer be ρ_1 and that of the lower layer be ρ_2, and let the (varying) depth of the top layer, in effect the depth of the Gulf Stream be D. Then geostrophic balance across the Gulf Stream, upper layer only results in

$$-fv = \frac{(\rho_2 - \rho_1)}{\rho_1} g \frac{\partial D}{\partial x}.$$

The phrase *reduced gravity* (g') is reserved for the expression

$$\frac{(\rho_2 - \rho_1)}{\rho_1} g$$

whence the northerly component of velocity v is given by the expression

$$v = \frac{g'}{f} \frac{\partial D}{\partial x}.$$

If the vorticity associated with this quantity is of the same order as the vorticity associated with the local value of the Coriolis parameter, then we can use dimensional analysis to deduce the length scale associated with the width of the Gulf Stream. Call this length scale L, then

$$f \sim \frac{g'}{f} \frac{D_1}{L^2}$$

since the vorticity associated with v is

$$\frac{\partial v}{\partial x} = \frac{\partial}{\partial x} \left(\frac{g'}{f} \frac{\partial D}{\partial x} \right).$$

Thus

$$L = \frac{\sqrt{g' D_1}}{f}$$

and the constant L is termed the *Rossby radius of deformation* which we will meet again later in Chapter 6 on waves. The value D_1 is representative of D and is the value it takes at the edge of the Gulf Stream. The width of the Gulf Stream is therefore L the value of which is deduced by using the values:

$$D_1 = 800m, \ f = 10^{-4}s^{-1}, \ \frac{(\rho_2 - \rho_1)}{\rho_1} = 2 \times 10^{-3}, \text{ and } g = 10ms^{-2}$$

giving $L = 40km$. This is a similar value as before but without the use of a perhaps controversial value of the eddy viscosity. To see how a solution to this model can be obtained, we proceed as follows. At the outer edge of the Gulf Stream where the fluid vorticity $\frac{\partial v}{\partial x}$ is negligibly small the depth of the Gulf

Stream is D_1. The mathematical statement of the conservation of potential vorticity thus gives

$$\frac{\zeta + f}{H} = \frac{\frac{\partial}{\partial x}\left(\frac{g'}{f}\frac{\partial D}{\partial x}\right) + f}{D} = \frac{f}{D_1} = \text{constant}.$$

Now in this model, $f = $ constant, and vorticity is conserved by the variation in the depth of the Gulf Stream balancing the effects of a northwards flow that varies east-west, across its direction. The equation for D is thus

$$\frac{\partial^2 D}{\partial x^2} = \frac{(D - D_1)}{L}$$

which is easily solved, but a solution is not attempted as the emphasis here is on dimensional analysis. The problem with the Stommel model is that it does not reproduce a counter current (the Munk model does). Thus we have two models, both idealised but both with shortcomings. This is quite typical, and serves to show how modelling operates and convinces you, if convincing you need that modelling is by no means an exact science.

2.6.3 El Niño and Coastal Upwelling

In the past five or so years, the phrase El Niño has entered the general vocabulary. Before this it was only known to the specialist oceanographic community, but now it seems any vagaries against the normal run of weather is routinely if sometimes wrongly blamed on "El Niño". Those who remember the fifties might recall nuclear tests being blamed similarly. In this case study, the aim is to describe a model of coastal upwelling. However in order to do so, it is appropriate today to embed this model in a general description of El Niño. The book by Philander (1990) should be consulted for an in depth study, what is given here is a brief outline of El Niño. The phenomenon of El Niño is part of ENSO (El Niño Southern Oscillation) which is a global climate idiosyncrasy. ENSO is the acronym given to an oscillation of the climate whereby one year there is warmer than usual eastern Pacific water in the region of New Guinea which in turn gives rise to a strong equatorial undercurrent running parallel to the equator from east to west at around 10°S. It is this current, warmer than surrounding seas that causes the problems. It reaches the coast of South America, Ecuador and Peru, destroys the nutrient rich upwelling (a model for which follows below) changes the climate causing sea fogs and coastal rain and is generally bad news. In other years, no such enhanced sea temperature in the east Pacific exists and so there is no destructive current destroying the coastal upwelling of South America. This leaves the rich cold waters to enable a thriving anchoveta fishing industry to aid the weak third world economies of Ecuador and Peru. In these years of more frequent El Niños, the opposite state is termed La Niña - it used to be merely normality. This interannual oscillation is now attracting a great deal of attention and research funding. Of course it is vital to understand what drives the phenomenon that devastates the economies of

Peru and Ecuador, but El Niño is credited with global climate disasters such as droughts, failure of monsoons (one of which lead to the Malaysian smog of 1998 - the burning of stubble, never a very good idea in that part of the world went out of control when the rains failed to materialise) and floods (USA 1997, and perhaps Mozambique 2000).

Unfortunately, the equatorial region is difficult to model, but lets try to describe the basics. One of the essential ingredients of a successful equatorial model is the correct dynamic modelling of the equatorial undercurrent. This is most easily described using local cartesian co-ordinates but valid across the equator. Such a co-ordinate system uses an equatorial β-plane approximation whereby the Coriolis parameter f is linearly dependent on the distance away from the equator y:

$$f = \beta y,$$

where $\beta = 2\Omega/R$, $\Omega = 7.29 \times 10^{-5} s^{-1}$ is the angular speed of the earth and $R = 6.36 \times 10^6 m$ is the radius of the earth. Another essential ingredient is local temperature. This can be incorporated through a warm upper layer that carries the current which enables us to a first approximation to use a two layer sea. The upper layer is a mere $100m$ deep and the lower layer is $2700m$ and represents the deep ocean down to the sea bed. Any stresses that may be present at the interface (interpreted as the thermocline) are ignored. The eastward flowing current in the upper layer can contain Rossby waves as f varies, but they have a special character, equatorial Rossby waves. The three equations valid for this model are:

$$\frac{\partial u}{\partial t} - fv = -g'\frac{\partial \eta}{\partial x} + \frac{\tau^x}{H},$$

$$\frac{\partial v}{\partial t} + fu = -g'\frac{\partial \eta}{\partial y} + \frac{\tau^y}{H},$$

$$g'\frac{\partial \eta}{\partial t} + c^2\left(\frac{\partial u}{\partial x} + \frac{\partial v}{\partial y}\right) = 0.$$

The first two equations represent quasi-geostrophic balance. The non-linear terms are neglected, and the equations have been integrated through the upper $100m$ layer, called H here and η is the sea surface elevation. The surface wind stress is (τ^x, τ^y) and becomes a body force because of the integration process. It will be noticed that reduced gravity g' rather than g appears due to the presence of the lower layer, and c is the baroclinic wave speed:

$$g' = \frac{\rho_2 - \rho_1}{\rho_1}g, \qquad c^2 = g'H.$$

These equations can be solved using the numerical methods of the next chapter, however further simplifications are possible once it is assumed that we wish to model an eastward equatorial jet. Of course formal elimination of the variables η and u is possible to leave the following single rather complicated equation for v.

$$\frac{\partial}{\partial t}\left(\frac{\partial^2 u}{\partial x^2} + \frac{\partial^2 v}{\partial y^2}\right) + \beta\frac{\partial v}{\partial x} - \frac{1}{c^2}\frac{\partial^3 v}{\partial t^3} - \frac{f^2}{c^2}\frac{\partial v}{\partial t} = F$$

where F is the forcing function

$$F = \frac{f^2}{c^2}\left[\frac{\partial}{\partial t}\left(\frac{\tau^x}{H}\right)\right] - \frac{1}{c^2}\left[\frac{\partial^2}{\partial t^2}\left(\frac{\tau^y}{H}\right)\right] + \frac{1}{H}\frac{\partial}{\partial x}\left(\frac{\partial \tau^x}{\partial y} - \frac{\partial \tau^y}{\partial x}\right).$$

This formal elimination can be quite valuable but must never be the equation eventually solved numerically. In performing the elimination process, extra derivatives are inevitably introduced and extra derivatives are the worst enemy of the numerical analyst. The value of the elimination lies in recognising dynamic balances. The equation for v is a spatial derivative of potential vorticity conservation. Sufficiently far from the equator and if the time scales are correct, this can be interpreted in potential vorticity form:

$$\frac{\partial}{\partial t}\left(\frac{\partial^2\psi}{\partial x^2} + \frac{\partial^2\psi}{\partial y^2} - \frac{f^2}{c^2}\psi\right) + \beta\frac{\partial\psi}{\partial x} = \left(\frac{\partial\tau^x}{\partial y} - \frac{\partial\tau^y}{\partial x}\right),$$

where

$$u = -\frac{\partial\psi}{\partial y} \qquad v = \frac{\partial\psi}{\partial x}.$$

Hence the extra contribution from the presence of the deep layer the third term on the left can be clearly seen. For much more on such equatorial models, consult Philander (1990). It turns out that in order to model the El Niño phenomenon one more ingredient is necessary, that is a y dependent eastward current. This is in order to conserve potential vorticity (the further from the equator, the greater the magnitude of f hence the smaller the fluid vorticity). The equations are thus:

$$\frac{\partial u}{\partial t} + U\frac{\partial u}{\partial x} - \left(f - \frac{\partial U}{\partial y}\right)v = -g'\frac{\partial\eta}{\partial x},$$

$$\frac{\partial v}{\partial t} + U\frac{\partial v}{\partial y} + fu = -g'\frac{\partial\eta}{\partial y},$$

$$\frac{\partial\eta}{\partial t} + U\frac{\partial\eta}{\partial x} + \frac{\partial}{\partial x}(Hu) + \frac{\partial}{\partial y}(Hv) = 0.$$

The depth of the mixed layer H is dependent on thermodynamics. It is recognised that some thermodynamics is essential in a model of El Niño, and it transpires that there is just enough in this model. We shall leave El Niño modelling here, there is much more in the book by Philander (1990) and more recent journal articles, however be warned. Articles and books on El Niño tend to the two extremes; either very technical because it is a tricky and not fully understood phenomenon, or too journalistic capitalising on media popularity to sell copy.

What we are interested in are coasts, and the effects of boundaries in general. In fact, no matter where the model is to be applied, the effects of the boundary will gradually migrate in. This is why Chapter 4 is devoted entirely to how boundary conditions are treated. Initially, if attention is focussed on an equatorial jet of the type rampant in El Niño years we have every right to assume that variations in the x direction are a lot less than those in the y direction. The dominant balance for v is

$$v = -\frac{\tau^x}{fH}$$

which arises from Ekman surface layer dynamics. At the equator itself, this of course breaks down due to the zero value of the Coriolis parameter f. The more accurate equation obeyed by v taking into account variations in y and possible inertial waves but ignoring derivatives with respect to x is

$$\frac{\partial^2 v}{\partial t^2} + f^2 v - c^2 \frac{\partial^2 v}{\partial y^2} = -\frac{f\tau^x}{H}.$$

It is all too easy to be over awed by the complexity of these models. It is true that they have to be three dimensional as thermal effects are central and waves must be present hence there is time dependence too. However, the message we wish to confer here is that there are special time and space scales which can be easily derived. Near the equator, $f = \beta y$ which in dimensional terms is βL where L is the horizontal length scale. If there are waves on this equatorial jet which are propagating with speed c then the two scales emerge: the distance $\sqrt{c/\beta}$ is the equatorial radius of deformation and the time scale $1/\sqrt{\beta c}$ determines the relative importance of the first two terms in the last equation for the speed of the jet v. These scales are larger than their counterparts in mid latitudes on a constant f model. The scale $\sqrt{c/\beta}$ is typically $250km$ (ten times a similar mid latitude value) and the time scale $1/\sqrt{\beta c}$ is 1.5days which is the local inertial time a distance $\sqrt{c/\beta}$ from the equator.

At the equator itself there are upwelling zones. These can be predicted using specialist equatorial models, but this is not done here. At coasts there are places where upwelling is both important economically and interesting physically. In fact 90% of all the worlds fish are caught in these biologically productive coastal upwelling zones. The deduction that an alongshore wind can produce an offshore Ekman drift and hence induce upwelling has already been made. Although there is little doubt that this Ekman mechanism provides an important source for upwelling there are features that this simple model is unable to reproduce. Amongst these are the equatorwards jet and the polewards undercurrent. One modification that we might try therefore is the introduction of stratification which is most easily done via a two layer model. This embodies an upper mixed layer, bounded below by a thermocline and below which is a cool deep layer extending to the sea bed. Another feature that the simple Ekman model cannot incorporate is waves. These are not considered here, but appear in Chapter 6. Waves propagate and their presence indicates the possibility of non-local forcing. This is important. Wind may force a wave which at the location of the forcing may not be significant, but at some distance could have been amplified by topography or coast to produce a major upwelling event. The two layer equations are quasi-geostrophic:

$$\frac{\partial u_1}{\partial t} - f v_1 = -g\frac{\partial \eta}{\partial x} + \frac{\tau^x}{\rho_1 H_1}$$

$$\frac{\partial v_1}{\partial t} + f u_1 = -g \frac{\partial \eta}{\partial y} + \frac{\tau^y}{\rho_1 H_1}$$

for the upper layer, and

$$\frac{\partial u_2}{\partial t} - f v_2 = -g \frac{\partial \eta}{\partial x} - g' \frac{\partial \eta_1}{\partial x}$$

$$\frac{\partial v_2}{\partial t} + f u_2 = -g \frac{\partial \eta}{\partial y} - g' \frac{\partial \eta_1}{\partial y}$$

for the lower layer. η_1 is the upward displacement of the interface (thermocline), H_1 the undisturbed depth of the upper layer and ρ_1 its density. Subtracting these equations, we get the following equations obeyed by $\hat{u} = u_1 - u_2$, $\hat{v} = v_1 - v_2$:

$$\frac{\partial \hat{u}}{\partial t} - f \hat{v} = -g' \frac{\partial \eta_1}{\partial x} + \frac{\tau^x}{\rho_1 H_1}$$

$$\frac{\partial \hat{v}}{\partial t} + f \hat{u} = -g' \frac{\partial \eta_1}{\partial y} + \frac{\tau^y}{\rho_1 H_1}.$$

More details of the underlying assumptions behind this model can be found in Gill(1982) (which, incidentally ranks amongst the best major texts in the physics of the ocean atmosphere system). Here let us concentrate on interpreting the model and its solution. The equation for each layer are forced unsteady geostrophic balances. The wind stress appears as a body force in the upper layer reflecting vertical integration (or equivalently the independence of variables on z the vertical co-ordinate). The forcing in the lower layer is due solely to interfacial slope. The variables $(\hat{u}, \hat{v})$ are the second (baroclinic) mode of velocity. The barotropic mode is not important but is $((H_1 u_1 + H_2 u_2)/(H_1 + H_2), (H_1 v_1 + H_2 v_2)/(H_1 + H_2))$ for those interested, where H_2 is the lower layer depth. It may seem odd that a model of upwelling, a velocity in the z direction, effectively neglects motion in this direction! This is a downside of using a simple model. However, upwelling *is* indicated through a rising of the thermocline which in this model means η_1 increasing. To progress, near a coast it is safe to assume that variations with respect to y along the coast are much smaller than variations with respect to $-x$, away from the coast. Note here that we continue to use x-East and y-North (unlike Gill (1982) - be warned) so that at an east coast x is into and not out from the coast. This explains the minus sign above. A local solution to the equations for $(\hat{u}, \hat{v})$ η_1 valid very close to this coast at $x = 0$ ($x < 0$ but very small) is

$$\hat{u} = \frac{\tau^y}{\rho f H_1}(1 - e^{xf/c_1})$$

$$\eta_1 = \frac{c \tau^y}{\rho g' H_1} e^{xf/c_1} t$$

$$\hat{v} = \frac{\tau^y}{\rho H_1} e^{xf/c_1} t$$

where
$$c_1{}^2 = \frac{g'H_1H_2}{H_1 + H_2}.$$

In order to derive this result, the third (continuity) equation needs to be used which takes the simplified form:

$$\frac{\partial \eta_1}{\partial t} + \frac{H_1 H_2}{H_1 + H_2} \frac{\partial \hat{u}}{\partial x} = 0.$$

From this solution it is immediately evident that η_1 grows with t which as we have said indicates upwelling in this model. In regions of upwelling the thermocline often rises enough to break the surface. This model does indeed also predict the equatorward current in the upper layer and the poleward undercurrent in the lower layer. The offshore flow is the steady Ekman solution and the flow parallel to the coast grows linearly with time. The interface also grows linearly with time. It indeed will eventually rise enough to break the surface and would thus seem to mimic reality. However, before this happens the solution will of course have become invalid as H_1 cannot be permitted to become zero. The length scale c_1/f is the baroclinic radius of deformation and is the appropriate dimensional estimate for the width of the coastal boundary layer. It is a lot less than the barotropic radius of deformation. Here are some values of parameters quoted by Gill (1982) from older published papers: $\tau^x = 0.1 Nm^{-2}$, $g' = 0.03 ms^{-2}$, $H_1 = 100m$ with $H_2 >> H_1$. Using these figures, $\hat{v}$ increases at a rate of $0.1 ms^{-1}$ per day, and the baroclinic radius of deformation is just over $10km$. The observed alongshore current speed is often greater than $0.2 ms^{-1}$ and the observed upwelling speed of order $10^{-4} ms^{-1}$. Both are consistent with this simple model which we therefore conclude captures the essential physics: wind stress driven; alongshore geostrophy; Ekman flow out from the coast. In reality there will be waves (see Chapter 6).

An alternative model is to use β-plane dynamics where the Coriolis parameter takes the form $f = f_0 + \beta y$. Most models that do this also include a variable depth, this is because the important dynamics is now the conservation of potential vorticity. An appropriate length scale for a coastal upwelling region with stratification, without explicitly performing any calculations is $\sqrt{c_1/\beta}$ which for the above parameter values is of the order of $400km$, an order of magnitude larger than the previous coastal upwelling model. Modelling upwelling using a β-plane model will thus involve the topography as well as barotropic time dependent effects (the Kelvin wave, see Chapter 6). The β effect also allows for Rossby (or planetary) waves which propagate energy westward hence providing a mechanism for the leakage of energy which prevents the building up of either equatorward current or poleward undercurrent. Gill (1982) still provides the best account of coastal upwelling, however there continue to be specialist articles, especially those involving the interaction of coastal upwelling with El Niño see for example Huyer (1987), Csanady (1997).

Chapter 3

Numerical Methods

3.1 Introduction

Numerical analysis is now a well established branch of mathematics. What we do not want to do here is to spend pages going through fundamental principles. However a brief recap is a good idea. For those who are familiar with finite differences the next section should be easy, however there are many environmental scientists without this background so no apology for the brief recap is necessary. Most of the effort should be channeled to understanding the wider implications of applying the methods and not so much on the fine details. It is accepted that what follows will not perhaps enable the reader to build his or her own model from equations to computer code, it should however help you to understand existing software and more importantly be better able to judge their pitfalls.

As we have seen from Chapter 2 the equations that describe the behaviour of the sea are differential equations; that is they involve rates of change of the basic quantities such as velocity, temperature, pressure, etc. Computers are digital which means that they cannot deal explicitly with continuous quantities. This was not a problem in the early 1960s and before when models were simple enough to be solved analytically by a variety of exact mathematical means. The use of exact mathematical methods still has its place in modelling and can bring much insight, however the use of the computer has opened up the modelling club to those who are not particularly mathematically inclined. In order to use the computer, the continuous quantities have to be converted into discrete quantities without unacceptable loss of accuracy. It is this step that is the province of numerical analysis. Numerical methods are the methods required to solve the discretised equations and we shall need to meet these too. It is also not obvious that the solutions to the resulting discrete equations will be the same or even close to the solutions of the exact differential equations.

There are three basic types of model used in oceanography. One is based on finite differences, one is based on finite elements and the third is based on

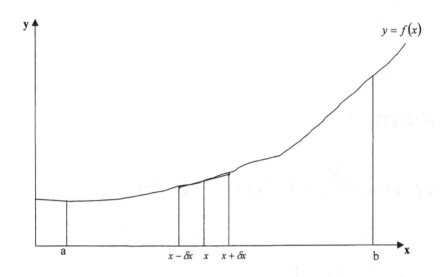

Figure 3.1: The graph of the function $f(x)$ and the three finite differences

spectral methods. The last method does not feature in this text and is really only used in environmental science for global meteorological modelling. Spectral methods certainly have little place in small scale oceanographic modelling.

Here is the basic idea behind finite differences. One wishes to replace a function $f(x)$ which is defined over a range, say $a \le x \le b$ by a finite series of values. If derivatives (i.e. rates of change) of $f(x)$ are required, then one of the three differences defined below can be used:-

1. forward difference:- $\Delta f = \dfrac{f(x + \delta x) - f(x)}{\delta x}$

2. centred difference:- $\delta f = \dfrac{f(x + \delta x) - f(x - \delta x)}{2\delta x}$

3. backward difference:- $\nabla f = \dfrac{f(x) - f(x - \delta x)}{\delta x}$.

These are shown graphically in Figure 3.1. Let us suppose that δx is the gap between each point on the x axis. This is called the step length and it is normal for this to remain the same throughout the calculation. This is because in adapting a quantity to be represented on a computer the step length is under our control and there is no earthly reason why it should not be constant for our convenience. Let us home in on one value of x in the range $a \le x \le b$ say x_i, let the point immediately to the left of this be x_{i-1} and the point immediately to the right be x_{i+1}. The above definitions thus become:-

1. forward difference:- $\Delta f = \dfrac{f(x_{i+1}) - f(x_i)}{\delta x}$

2. centred difference:- $\delta f = \dfrac{f(x_{i+1}) - f(x_{i-1})}{2\delta x}$

3. backward difference:- $\nabla f = \dfrac{f(x_i) - f(x_{i-1})}{\delta x}$

where we can write $\delta x = x_{i+1} - x_i$, the step length. It can be seen from Figure 3.1 that the three differences are approximations to the gradient of $f(x)$ at the point x_i. Closer examination reveals that the centred difference looks the best of the three approximations. However, this does depend on the behaviour of the function $f(x)$. In general the centred difference is the most accurate, our definition masks this as it is based on a double step length (in order to avoid half steps). It is the symmetric difference using information from both sides of x_i in order to approximate the gradient and so ought to be the most accurate. The forward difference has a special place as it *predicts*. If x is time, then the forward difference contains the value of f at the later time step, and an equation that contains a single time derivative may be able to be rearranged in order that this value is on the left whilst the right-hand side contains only values of quantities at the present time. As these are all known, we have a prediction for $f(x_{i+1})$. Perhaps not a very accurate one, but one nevertheless. In the next section we go into more detail about the methods used and how accurate they are.

3.2 Finite Differences and Error Analysis

Fine detail on the kind of numerical methods we will use can be found in specialist texts (for example G D Smith's Numerical Solution of Partial Differential Equations published by Oxford University Press). This kind of text will be concerned with the estimation of errors in a precise way, with convergence and with inconsistency. These subjects are at home in undergraduate mathematics courses, not here. Instead, what will be done here is a brief introduction to give the flavour of how finite difference methods can be applied to our kind of equations. Partial Differential Equations that contain a single time derivative of the first order (e.g. $\dfrac{\partial u}{\partial t}$) and no other time derivatives are of the type called parabolic partial differential equations. This is the case no matter how many or what kind of derivatives with respect to x, y or z occur. One of the simplest parabolic differential equations, and the one all the books tend to use to illustrate the action of numerical methods on parabolic differential equations is the diffusion equation. We shall meet this in chapter 5 in its own right, but for now we state it as follows:

$$\frac{\partial u}{\partial t} = \kappa \frac{\partial^2 u}{\partial x^2},$$

where κ is a constant called the diffusivity. If a forward difference in time but a centred difference in space is used, then at any particular point the discrete version of the diffusion equation is

$$\frac{u_i^{s+1} - u_i^s}{\Delta t} = \kappa \frac{u_{i+1}^s - 2u_i^s + u_{i-1}^s}{(\Delta x)^2}$$

where we have written

$$u(i\Delta x, s\Delta t) = u_i^s,$$

Δx being the step length for space and Δt that for time. The crucial point is that it is possible for us to make u_i^{s+1} the subject of an explicit formula, namely

$$u_i^{s+1} = u_i^s + r(u_{i+1}^s - 2u_i^s + u_{i-1}^s)$$

where

$$r = \frac{\kappa \Delta t}{(\Delta x)^2}.$$

When it is possible to give an explicit expression for the value of the variable at the next time step, the method is (naturally enough) called an *explicit* scheme or *explicit* numerical method. The reason all practical numerical schemes are not explicit is that unless r is less than a specific value (1/2 for the diffusion equation) the scheme does not work. This is because it oscillates instead of converging on a value close to that expected for $u(x, t)$. The name for this is instability, and there is no cure other than changing the finite difference scheme, either by making r smaller or by choosing a different scheme altogether. Making r smaller is often impractical as it would mean such a small time step that forecasting would be too inefficient. Imagine having to use a 30 second time step for a one day forecast, this would take 172800 steps! Not using an explicit method but instead using an alternative method that allows a 7 minute (say) time step would seem a better way forward. One example of this alternative method is that due to Crank and Nicolson (the Crank-Nicolson method). This discretises the diffusion equation as follows:

$$\frac{u_i^{s+1} - u_i^s}{\Delta t} = \frac{\kappa}{2} \left[\frac{u_{i+1}^{s+1} - 2u_i^{s+1} + u_{i-1}^{s+1}}{(\Delta x)^2} + \frac{u_{i+1}^s - 2u_i^s + u_{i-1}^s}{(\Delta x)^2} \right].$$

The problem is now obvious. The predicted (*unknown*) values of u at the later time step occur on both sides of the equation and at three different points in space. There is therefore no explicit single equation for u_i^{s+1}. Instead all the equations for all the discrete values of u have to be written down and a matrix equation set up and solved for all the values of u at the later time step. This is an example of an *implicit* scheme, but the inconvenience of having to invert matrices at each time step is out weighed by the method being unconditionally stable. Figure 3.2 gives a pictorial view of how the Crank-Nicolson method operates to solve the diffusion equation. Of course, there are many different schemes possible, just think of the generalisation

$$\frac{u_i^{s+1} - u_i^s}{\Delta t} = \kappa \left[\frac{\theta(u_{i+1}^{s+1} - 2u_i^{s+1} + u_{i-1}^{s+1})}{(\Delta x)^2} + \frac{(1-\theta)(u_{i+1}^s - 2u_i^s + u_{i-1}^s)}{(\Delta x)^2} \right],$$

where θ is a parameter which we the user are free to prescribe. The value $\theta = 1/2$ regains the Crank-Nicolson scheme. This generalisation is sometimes useful, but the Crank-Nicolson version is almost always best certainly for the

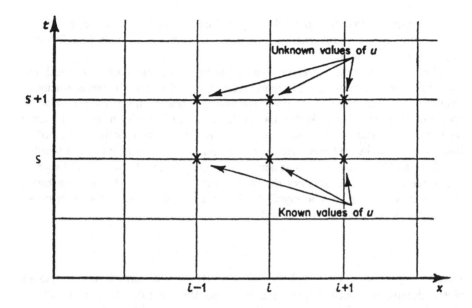

Figure 3.2: The Crank-Nicolson scheme drawn schematically

diffusion equation. If $\theta = 1$, then the scheme is called *fully implicit*. Fully implicit schemes have the merit of usually being the most stable. Unfortunately they are prone to be inaccurate and exhibit large truncation error (see the next paragraph for an explanation of this terminology). There is a general class of numerical schemes called Lax Wendroff schemes which are $O(\Delta t)^2$ in terms of truncation error which is achieved through the cancellation of the $O(\Delta t)$ generated by the forward difference in time with the same term generated by the finite difference of the spatial derivative. To illustrate this, consider the one dimensional advection equation:

$$\frac{\partial u}{\partial t} + c\frac{\partial u}{\partial x} = 0.$$

The finite difference approximation to this using a Lax-Wendroff scheme would be

$$\frac{u_i^{s+1} - u_i^s}{\Delta t} + c\frac{u_{i+i}^s - u_{i-1}^s}{2\Delta x} = \frac{c^2\Delta t}{2}\left(\frac{u_{i+1}^s - 2u_i^s + u_{i-1}^s}{(\Delta x)^2}\right).$$

The lowest order truncation error for this scheme seems to be due to the $O(\Delta t)$ error in the explicit finite time difference on the left. However this is exactly cancelled by the equivalent term on the right since

$$\frac{\Delta t}{2}\frac{\partial^2 u}{\partial t^2} = \frac{\Delta t}{2}\frac{\partial}{\partial t}\left(-c\frac{\partial u}{\partial x}\right) = \frac{c^2\Delta t}{2}\frac{\partial^2 u}{\partial x^2}.$$

Precisely how these errors are calculated involves knowledge of Taylor's series. More is said about this a little later in this section. Many fancy schemes with useful properties have now been developed, but the time has come to define the terms round-off error and truncation error more precisely. This is done formally in the next paragraph.

There are two types of error that occur in employing numerical methods. This discounts the third kind, human error; copying incorrectly, transposition of numbers, etc. The two errors we will consider are *round-off* error and *truncation* error and they are easy to distinguish in theory, but of course an error is an error whatever the cause. Round-off error occurs because infinite decimals cannot be held in even the largest computer. The computer representation of the number $\pi = 3.14159265\ldots$ has to stop at some point, even if the number itself does not. Of course, round-off error is not the problem it used to be with small machines, but it cannot be ignored altogether. As an example, it is common in ocean science to be required to compute the value

$$\frac{1}{\omega^2 - f^2}$$

where $\omega = 1.405 \times 10^{-4}$ and $f = 1.12 \times 10^{-4}$. The first is the frequency of the dominant (twice daily) lunar tide and the second a local value of the Coriolis parameter. Direct computation is of course possible, but undesirable as it involves the subtraction of two numbers that are very similar in magnitude. This would be prone to round-off error. Far better to compute

$$\frac{1}{((\omega/f)^2 - 1)f^2}$$

which would always give a more accurate answer. Round-off error is no longer a serious problem.

Truncation error comes about because a derivative has been approximated by a difference. It cannot be eliminated entirely but can be reduced by judicious choice of finite difference scheme. One crude way of reducing truncation error is to decrease the step length(s). This will virtually always give a more accurate answer, but at a price. The model will be slower to compute due to the increase in size of the number of unknowns. It will also take longer for any useful predictions to emerge as the time step will be shorter too. A more sophisticated way of reducing truncation error would be to change the scheme. As already mentioned, to those with a knowledge of Taylor's series or Taylor's theorem as it is alternatively known, it is quite easy to estimate the order of the truncation error for a given scheme. Let us see how this can work. (If Taylor's theorem is a deep mystery, skip the next bit.) Taylor's theorem for a function of a single variable $f(x)$ about the point $x = a$ takes the form

$$f(a + h) = f(a) + hf'(a) + \frac{h^2}{2!}f''(a) + \frac{h^3}{3!}f'''(a) + \ldots$$

where a dash denotes a derivative. This means that if $f'(a)$ is made the subject of this formula we can write

$$f'(a) = \frac{f(a + h) - f(a)}{h} - \frac{h}{2}f''(a) - \frac{h^2}{6}f'''(a) + \ldots.$$

Now the first term is simply the forward difference approximation for the first derivative of $f(x)$ at the point $x = a$. Therefore the term

$$\frac{h}{2} f''(a)$$

is the leading error term for this difference. In this way it is possible to estimate truncation error. The forward difference (and the backward difference) have truncation errors that are proportional to h. It is similarly shown (by subtraction of Taylor series) that the truncation error of a centred difference scheme is proportional to h^2 hence centred differences are in general more accurate. The accuracy is of course dependent upon the size of h, the step length as well as the behaviour of the derivatives of f. We are thus led to the unsurprising conclusion that approximating the derivatives of functions that vary rapidly leads to greater error than approximating the derivatives of smooth functions. In a realistic finite difference scheme, it is possible to use Taylor series expansions to estimate the leading term in the general truncation error for the whole scheme. This calculates what is termed the *local* truncation error. The *global* truncation error is the sum of all these local errors and is of course larger. Decreasing the step length in a scheme that is first order (truncation error $\sim h$) accurate will not decrease the global error as the more accurate value at each step will be exactly cancelled by the increase in steps. For a scheme that is accurate to $\sim h^2$ at each step the global error will be $\sim h$.

This may come as quite a surprise, but it is possible to select a finite difference approximation to a partial differential equation in all innocence, but for this approximation not to home in on the original equation once the step lengths have become infinitesimally small. The reason this can happen is amongst the error terms which are expressed as derivatives via Taylor's series (see above) there could be one or more that remains finite in the limiting process. There are many ways of letting the step lengths tend to zero, so you could be unlucky and choose a wrong one. When this happens, the scheme is called *inconsistent*. The only sure way of testing for consistency is to perform the Taylor's series expansions together with the limiting processes manually as it were. The schemes outlined here will all be consistent.

A popular phrase encountered in the numerical methods literature on finite difference techniques is *upwind differencing*. It is worth explaining what this means. Suppose a variable c is being modelled (it might be concentration, see Chapter 5) then the scheme whereby if the speed $u < 0$

$$\frac{\partial c}{\partial x} = \frac{c_i^{s-1} - c_{i-1}^{s-1}}{\Delta x}$$

but if the speed $u > 0$ then

$$\frac{\partial c}{\partial x} = \frac{c_{i+1}^{s-1} - c_i^{s-1}}{\Delta x}$$

is called upwind differencing. The phrase stems from that c can only be detected if one is downwind of it. The resulting scheme is stable, but the use of the one

sided differences means that there is large truncation error unless particular care
is taken in the design of the overall scheme. Upwind schemes feature prominently
in software packages of the type mentioned in section 3.6.

3.3   Applying finite differences to coastal sea modelling

In Chapter 6 the modelling of various aspects of continental shelf seas will be investigated, we thus postpone a detailed description of the physical environment
until then. In particular, the description will include an outline of waves and
the terms used to describe waves. It will be assumed here that the terms wavelength, amplitude and period are familiar. Those in need of a short refresher
course may like to turn straight away to Chapter 6 where waves are introduced
as if for the first time, and section 6.5 where tides on the continental shelf are
discussed. Here we state that the wavelength of a typical tide is 1000km and
its amplitude is typically less than a few metres. The wave slope is thus around
10^{-5}. This means that tides are extremely long waves when considered as water waves. It is therefore acceptable to assume that the surface is sinusoidal
in shape and to apply shallow water wave theory. Also, as the depth (around
200m on the continental shelf) is very much less than the wavelength, the tidal
currents at the surface will not have any room to decay before the sea bed.
This is because in linear water wave theory the currents decay with depth as
$e^{-z/\text{wavelength}}$ where z is the distance beneath the sea surface, which is virtually
equal to one. To a first approximation therefore, the tide in a continental shelf
can be modelled as a z independent sinusoidal oscillation. As the sea is also
approximately hydrostatic, the pressure $p(x, y, t)$ will obey the equation

$$\frac{\partial p}{\partial z} = -\rho g$$

from which upon integration with respect to z gives

$$p = \rho g(\eta - z)$$

where η is the elevation of the sea surface above mean sea level. If the hydrodynamic equations are considered, then the non-linear terms are ignored as are
variations with respect to z and the vertical component of current. What results is a pair of simple partial differential equations in the *three* unknowns u, v
the horizontal components of current and η. The third equation that closes the
system is given by the conservation of mass and is derived in Chapter 2 section
2.4.4. The resulting set of three equations in three unknowns is

$$\frac{\partial u}{\partial t} - fv = -g\frac{\partial \eta}{\partial x}$$

$$\frac{\partial v}{\partial t} + fu = -g\frac{\partial \eta}{\partial y}$$

$$\frac{\partial \eta}{\partial t} + \frac{\partial}{\partial x}(hu) + \frac{\partial}{\partial y}(hv) = 0$$

which are the Laplace Tidal Equations (sometimes abbreviated to LTE). Insight into the behaviour of the solutions to these equations in a tidal situation can be gained by assuming that the time dependence is sinusoidal, and although the principal aim of this chapter is finding numerical solutions let us proceed in this way before using any numerical methods. Thus it is assumed that

$$\eta(x,y,t) = A(x,y)e^{i\omega t}, \quad u(x,y,t) = U(x,y)e^{i\omega t}, \quad v(x,y,t) = V(x,y)e^{i\omega t}.$$

Substituting these into the first two of the Laplace Tidal Equations gives explicit expressions for $U(x,y)$ and $V(x,y)$ namely

$$U(x,y) = \frac{1}{\omega^2 - f^2}\left[i\omega g\frac{\partial A}{\partial x} + fg\frac{\partial A}{\partial y}\right]$$

$$V(x,y) = \frac{1}{\omega^2 - f^2}\left[i\omega g\frac{\partial A}{\partial y} - fg\frac{\partial A}{\partial x}\right],$$

or multiplying by the exponential factor $e^{i\omega t}$ to regain the original variables,

$$u(x,y,t) = \frac{1}{\omega^2 - f^2}\left[i\omega g\frac{\partial \eta}{\partial x} + fg\frac{\partial \eta}{\partial y}\right]$$

$$v(x,y,t) = \frac{1}{\omega^2 - f^2}\left[i\omega g\frac{\partial \eta}{\partial y} - fg\frac{\partial \eta}{\partial x}\right].$$

The presence of the complex unit $i = \sqrt{-1}$ is confusing for some. It need not be. It represents that there are phase differences between the variables. In fact, substituting for u and v into the third of the Laplace Tidal Equations gives a single equation for η (or equivalently its amplitude A). This equation is

$$\frac{(\omega^2 - f^2)}{g}\eta + \left[\frac{\partial}{\partial x}\left(h\frac{\partial \eta}{\partial x}\right)\right] + \left[\frac{\partial}{\partial y}\left(h\frac{\partial \eta}{\partial y}\right)\right] + \frac{if}{\omega}\left|\frac{\partial(h,\eta)}{\partial(x,y)}\right| = 0.$$

The last expression $\left|\dfrac{\partial(h,\eta)}{\partial(x,y)}\right|$ is the Jacobian of the parameter h and the variable η defined by the determinant

$$\left|\frac{\partial(h,\eta)}{\partial(x,y)}\right| = \left|\begin{array}{cc} \frac{\partial \eta}{\partial x} & \frac{\partial \eta}{\partial y} \\ \frac{\partial h}{\partial x} & \frac{\partial h}{\partial y} \end{array}\right|.$$

This expression is zero if the depth is constant. For the mathematically minded we note that the complete story is as follows: provided neither h nor η are identically zero the Jacobian term is only zero if there is a functional relationship between the parameter h and the variable η which of course can never be the case. In the idealised case $h = $ constant the equation obeyed by η is the Helmholz wave equation

$$\nabla^2\eta + \lambda^2\eta = 0$$

where we have used the standard notation

$$\nabla^2 \equiv \frac{\partial^2}{\partial x^2} + \frac{\partial^2}{\partial y^2}, \text{ and } \lambda^2 = \frac{\omega^2 - f^2}{gh}.$$

Either of these equations are solved together with boundary conditions which are that there is no flow across coasts and that the value of η is prescribed at open boundaries.

In reality the modelling of tides has to include the effects of the sea-bed in terms of frictional effects. These produce a z dependence and complicate the equations. The details of this must wait until the next chapter, instead here the focus will be on the kind of numerical schemes that might be used in practical tidal models of particular seas. In this chapter so far we have introduced some numerical methods that are useful for the modelling of coastal seas. Explicit methods are seldom used nowadays, but basic ideas and concepts are conveniently introduced through them. Their principal drawback is the time step has to be inordinately small to preserve stability. This has already been said in the last section. For example, a time explicit scheme based on centred space differences for the Laplace Tidal Equations introduced above would be:

$$u_{i,j}^{s+1} = u_{i,j}^s + f\Delta t v_{i,j}^s - \frac{g\Delta t}{2\Delta x}(\eta_{i+1,j}^s - \eta_{i-1,j}^s)$$

$$v_{i,j}^{s+1} = v_{i,j}^s - f\Delta t u_{i,j}^s - \frac{g\Delta t}{2\Delta y}(\eta_{i,j+1}^s - \eta_{i,j-1}^s)$$

$$\eta_{i,j}^{s+1} = \eta_{i,j}^s - \frac{\Delta t}{2\Delta x}(\eta_{i+1,j}^s h_{i+1,j} - \eta_{i-1,j}^s h_{i-1,j})$$

$$- \frac{\Delta t}{2\Delta y}(\eta_{i,j+1}^s h_{i,j+1} - \eta_{i,j-1}^s h_{i,j-1}).$$

However if you try to solve these on a computer they will be unstable no matter how small the time step. We will return to this particular set of finite difference equations after making some more general statements.

It can be proved that the error due to the round off error and truncation error separately satisfy the chosen finite difference representation of the partial differential equations. In order to ascertain whether or not the scheme is stable, a common procedure is to use a method based on Fourier analysis. This means assuming that the error $\epsilon(x, t)$ behaves like

$$\epsilon(x, t) = e^{at}e^{ik_n x}$$

where we have assumed a one dimensional (x, t) equation for simplicity. Substituting into the finite difference representation results in an expression for e^{at} in terms of $k_n \Delta x$ and Δt where Δx and Δt are, of course the time and space step lengths respectively. Now, if the original partial differential equation contains solutions that are wave like with wave celerity or wave speed c (see Chapter 6) then it is possible to define a non dimensional number called the Courant number C where

$$C = c\frac{\Delta t}{\Delta x}$$

and for stability it is essential for this scheme that $C \leq 1$. This expresses that the waves naturally occurring in the physics represented by the original equation must not travel faster than the "numerical wave speed". If this happens, the real waves cannot be captured by the numerical scheme and instability results. This inequality is called the Courant-Friedrichs-Lewy condition or CFL condition for short. It is a necessary condition for the stability of any given scheme. Sadly, it is by no means sufficient! The CFL condition is of course different for other schemes. A general interpretation of the CFL condition is that the numerical domain of dependence of a finite difference scheme must include the domain of dependence of the associated partial differential equation. For schemes that involve a large number of points the CFL condition can be hard to derive. More advanced books on numerical methods, e.g. Durran (1999) should be consulted for more information.

For the explicit scheme discretising the Laplace Tidal Equations, substituting the Fourier expressions

$$\eta_{i,j}^s = a e^{\alpha t} e^{i(k_m x + l_m y)}$$
$$u_{i,j}^s = b e^{\alpha t} e^{i(k_m x + l_m y)}$$
$$v_{i,j}^s = c e^{\alpha t} e^{i(k_m x + l_m y)}$$

into the finite difference equations leads to an imaginary value for α. We deduce that the solution thus does not settle down. The inclusion of friction in the model helps it to be stable, but the truncation error and the friction term can be of the same order which is inconvenient to say the least.

3.4 The Arakawa Grids

The history of the numerical modelling of ocean currents is not very long - it dates from the mid-1960s - but several types of finite difference grid are in use and we take this opportunity to introduce the main varieties. For the moment we will only consider the horizontal discretisation of the variables u, v and η, which are the eastward current, the northward current and the surface elevation, respectively. There are in fact three grids (that is, ways of discretising these variables in the horizontal) still in use, and these are shown diagrammatically in Figure 3.3; (a) is the Arakawa A grid, (b) is the Arakawa B grid, and (c) is the Arakawa C grid. In the A grid, all variables are evaluated at the same location. At first sight this may seem logical, but bearing in mind that u and v are related to gradients of η, this turns out not to be very convenient. The B and C grids were developed so that points where the elevation was evaluated were always *between* points where the current was evaluated. This was first done by A. Arakawa (1966) who was a meteorological modeller. In the B grid, shown in Figure 3.3(b), both u and v are evaluated at the same point and the velocity points are situated at the point that is equidistant from the four nearest elevation points. In the C grid, this is not the case. Instead, the u points lie east and west of η points, and the v points lie north and south of the η points. This is

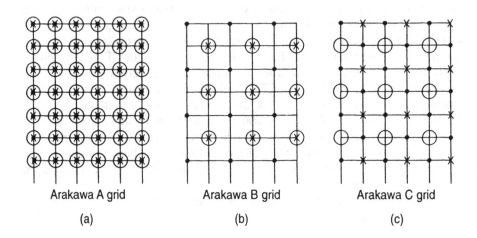

Figure 3.3: The three Arakawa grids. · elevation (η), ◯ northwards current (v),
× eastwards current (u).

shown in Figure 3.3(c), and it is this Arakawa C grid that is most popular today
with ocean and continental shelf modellers. However, the B grid does have the
advantage of allowing a semi-implicit representation of the Coriolis terms and
is also very useful for certain coarse grid schemes. It is also more stable under
certain circumstances. Other grids are of course possible; in particular, it is
more accurate to involve more values than just those on the neighbouring grid
points. In practice it is found that much the same increase in accuracy can be
achieved merely by decreasing the step size. This is also much more convenient
than complicating the difference equations by the inclusion of many more terms.

   In the vertical, a common device is to use σ-coordinates. This is the name
given to the co-ordinate system whereby σ replaces z through the formula:

$$\sigma = \frac{\eta - z}{h + \eta},$$

so that the sea surface $z = \eta$ is at $\sigma = 0$ and the sea bed $z = -h$ is at
$\sigma = 1$. In this fashion the domain of the problem (in the vertical) is flat at
the top and bottom. The down side of using the σ-coordinate system is that
derivatives of z have to be transformed into derivatives in terms of σ, which are
more complicated. The sea bed and surface boundary conditions are, however,
greatly simplified.

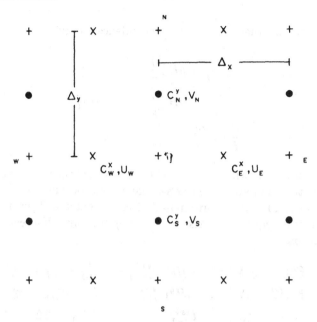

Figure 3.4: Richardson lattice: spatial arrangement of coefficients and variables. + pressure (η); × U-component; • V-component.

3.5 Case Studies

3.5.1 Semi-implicit methods

This first example is loosely based on a paper by Backhaus (1983). It is essentially a two dimensional (horizontal) model and what follows is a description of the discretisation in finite difference form. The grid shown in Figure 3.4 is the fully staggered grid, already introduced as the Arakawa C grid. Here it is called the Richardson grid. It will be remembered that there are three kinds of points, denoted by plus symbols, cross symbols and blob symbols. Surface elevation is evaluated at plus points. The differential equations obeyed by a sea have already been stated. The same Laplace Tidal Equations are used here, but with the addition of forcing terms. The equations are written in terms of vertically integrated transport (U, V) defined by

$$U = \int_{-h}^{\eta} u\, dz$$

$$V = \int_{-h}^{\eta} v\, dz$$

whence the equations obeyed by (U, V) are obtained by integrating the first two of the Laplace Tidal Equations with respect to z through the vertical (the third equation is merely rewritten, with Hu and Hv replaced by the above expressions for U and V as now there may be a z dependence) to give:

$$\frac{\partial U}{\partial t} + gH\frac{\partial \eta}{\partial x} = X - \tau_B^x$$

$$\frac{\partial V}{\partial t} + gH\frac{\partial \eta}{\partial y} = Y - \tau_B^y$$

$$\frac{\partial \eta}{\partial t} + \frac{\partial U}{\partial x} + \frac{\partial V}{\partial y} = 0.$$

The more astute of you will have noticed the absence of the Coriolis terms. These and perhaps the wind stress and non-linear terms too have been incorporated into the terms (X, Y) which have materialised as "catch-all" terms on the right of the momentum equations. Also $H = h + \eta$. In finite difference form, these equations become

$$U_{i,j}^{s+1} = F^x(U_{i,j}^s + \Delta t(X_{i,j}^s - gH^x(\eta_{i+1,j}^{s+1} - \eta_{i-1,j}^{s+1} + \eta_{i,j+1}^s - \eta_{i,j-1}^s)/2\Delta x))$$

$$V_{i,j}^{s+1} = F^y(V_{i,j}^s + \Delta t(Y_{i,j}^s - gH^y(\eta_{i+1,j}^{s+1} - \eta_{i-1,j}^{s+1} + \eta_{i,j+1}^s - \eta_{i,j-1}^s)/2\Delta y))$$

$$\eta_{i,j}^{s+1} = \eta_{i,j}^s - \Delta t((U_{i+1,j}^{s+1} - U_{i-1,j}^{s+1} + U_{i+1,j}^s - U_{i-1,j}^s)/2\Delta x$$

$$+ \ (V_{i,j+1}^{s+1} - V_{i,j-1}^{s+1} + V_{i,j+1}^s - V_{i,j-1}^s)/2\Delta y).$$

Finite difference equations always look more complicated than they are, and these are certainly no exception! Backhaus favours the spatial centred difference whereby for example

$$\frac{\partial U}{\partial x} \approx \frac{U_{i+1,j}^s - U_{i-1,j}^s}{2\Delta x}.$$

The normal notation whereby the value of U at the point where $x = i\Delta x$, $y = j\Delta y$ at time $s\Delta t$ is denoted by $U_{i,j}^s$ is used. The approximation used for the first terms of each equation, $(\partial \eta/\partial t$ etc.), is a forward difference. The whole basis of using finite differences as an approximation to the equations is to be able to predict what will happen to the variables u, v and η given what has and is happening to them. This means, as far as the mathematics is concerned, that we need to use forward differences whenever possible to discretise the time rates of change. However, it will be noticed that in this scheme, later times appear on the right as well as on the left. This is therefore an *implicit* scheme which needs matrix algebra for solution. The bottom stress term (τ_B^x, τ_B^y) was expressed in terms of the square of the of the local velocity, the standard quadratic friction law, and in the finite difference equations it is the terms F^x, F^y that are a non-dimensional friction function which arises from using a semi-implicit form for this bottom stress (τ_B^x, τ_B^y). By semi- implicit, we mean a Crank-Nicolson type of discretisation, half at this time, half at a later time as can be seen in the rest of the finite difference equations. In his paper, Backhaus uses this scheme to compute the depth mean currents in the Kattegat (between Denmark and Sweden). We have yet to say anything about satisfying boundary conditions. This general area is extremely important and the subject of the next chapter.

3.5.2 A Three Dimensional Finite Difference Scheme

In the body of this chapter, some details of how finite difference schemes can be applied to shallow water dynamics have been given. Here we give an example of modelling in which the equations are fully three dimensional and the vertical variation is catered for by a Galerkin weighted residual technique. By its very nature, this gets quite mathematical, so if this is not to your taste, move on to the next subsection.

The fully three dimensional equations that need to be solved are:

$$
\frac{\partial u}{\partial t} + u\frac{\partial u}{\partial x} + v\frac{\partial u}{\partial y} + w\frac{\partial u}{\partial z} - fv = -g\frac{\partial \eta}{\partial x} + \frac{\partial}{\partial z}\left(N_z\frac{\partial u}{\partial z}\right)
$$
$$
+ \frac{\partial}{\partial x}\left(N_x\frac{\partial u}{\partial x}\right) + \frac{\partial}{\partial y}\left(N_y\frac{\partial u}{\partial y}\right),
$$

$$
\frac{\partial v}{\partial t} + u\frac{\partial v}{\partial x} + v\frac{\partial v}{\partial y} + w\frac{\partial v}{\partial z} + fu = -g\frac{\partial \eta}{\partial y} + \frac{\partial}{\partial z}\left(N_z\frac{\partial v}{\partial z}\right)
$$
$$
+ \frac{\partial}{\partial x}\left(N_x\frac{\partial v}{\partial x}\right) + \frac{\partial}{\partial y}\left(N_y\frac{\partial v}{\partial y}\right),
$$

$$
\frac{\partial \eta}{\partial t} + \frac{\partial}{\partial x}(\int_{-h}^{\eta} u\,dz) + \frac{\partial}{\partial y}(\int_{-h}^{\eta} v\,dz) = 0,
$$

where the non-linear terms have been included using a cartesian (x, y, z) local set of axes. Friction takes the form of eddy-viscosities N_z vertically and N_x and N_y horizontally. This model for friction was in common use before the more sophisticated turbulence closure schemes (see later) appeared. First of all, the variation with depth needs to be dealt with, and the first step is the introduction of a slightly different but equally valid sigma co-ordinate defined this time by

$$
\sigma = \frac{z+h}{h+\eta}.
$$

This has the effect of transforming the vertical co-ordinate so that the new domain lies between $\sigma = 1$ the sea surface and $\sigma = 0$ the sea bed. The next step is the introduction of vertical "modes":

$$
u(x, y, \sigma, t) = \sum_{p=1}^{M} A_p(x, y, t)F_p(\sigma)
$$

$$
v(x, y, \sigma, t) = \sum_{p=1}^{M} B_p(x, y, t)F_p(\sigma).
$$

These are now substituted into the three equations of motion which is straightforward but demands accurate algebraic skills. The result is not given here, but can be found in for example Wolf (1982). The next step is to discretise the formidable set of equations! The favoured finite difference method is to use an ADI (alternating direction implicit) technique. The terms involving η, and the

non-linear terms in both the momentum equations and the continuity equations
are treated implicitly, the rest are treated explicitly. Again, full technical details
are not given, but there is evidence of some instability (Wolf (1982)). It is just
that the instability grows very slowly, partially damped by the friction present
and does not interfere unduly with the solution. The vertical eddy viscosity
terms are in fact split into two (the DuFort-Frankel method, see books on finite
difference schemes for fluid flow, e.g. Hirsch (1988)). The resulting matrix turns
out to be tri-diagonal which is conveniently solved by using a double sweep algorithm (called the Thomas algorithm, see Roache (1976)). The scheme was
applied to the prediction of tides in an idealised rectangular bay and then to
predict the tides in the Bristol channel in the UK. The cotidal chart (theoretical)
and results from the ADI scheme for the rectangular bay are depicted in Figure
3.5. When the time step was tripled, there was evidence of frictional effects,
but the general shape of the cotidal and corange lines was still recognisable.
According to the CFL criteria, the time step Δt must be chosen so that

$$\Delta t < \frac{\Delta x}{\sqrt{2gh}} = 1256s.$$

The chosen time step ($1200s$) is inside this, but experimenting with tripling
it takes the scheme outside the CFL condition. This also accounts for the
poorer resolution. The agreement between the tide in a rectangular gulf and
this numerical model must certainly be considered to be very good. Wolf (1982)
also undertook experiments with wind driven flow. The southwest detail of the
model was driven by a wind stress and as expected, the dominant oscillation
was driven by the basin width. These are called *seiches* and represent a sloshing
mode with period given by Merian's formula:

$$\frac{2L}{\sqrt{gh}},$$

where L is the width of the basin. This period here is $18 - 20$hrs. and was
well simulated by the ADI scheme. More details about seiches are given in
Proudman (1953), and in Leblond and Mysak (1978).

Turning to the Bristol channel, this is notoriously difficult to model not least
because of the complex geometry and the drying regions. Two dimensional models fail here. It is a severe test for all three dimensional models, nevertheless
the following table shows that the overall agreement with observed tide is quite
good. The eddy viscosity used here has a parabolic vertical profile. One of
the advantages with this particular model was the ability to tune it using the
eddy viscosity, and the paper by Wolf (1982) gives a value of equivalent bottom
friction co-efficient 0.001. With a value of 0.004, the depth had to be increased
everywhere by $5m$ which removed all the drying regions before there was reasonable agreement with observation.

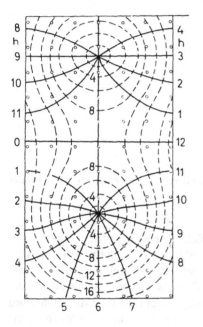

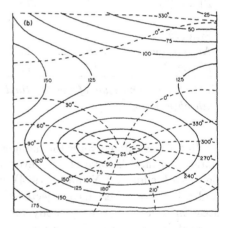

Figure 3.5: The cotidal and corange lines in a rectangular bay using (a) theory and (b) the ADI scheme with a time step of 1200s.

	Observed	Model k=0.001
Avonmouth		
Amplitude (cm)	422	429
Phase (°)	202	202
Newport		
Amplitude (cm)	413	363
Phase (°)	198	199
Flat Holm		
Amplitude (cm)	390	412
Phase (°)	192	192
Port Talbot		
Amplitude (cm)	315	319
Phase (°)	173	178
Swansea		
Amplitude	315	317
Phase (°)	173	178

Table 3.1 Comparison of the M_2 tidal amplitudes and phases from BCF model with observations.

In the past fifteen years the manner in which the effects of friction are incorporated in models have become ever more sophisticated. The pioneers in this field are Mellor and Yamada (1974) and Blumberg and Mellor (1987). These and other researchers have rejected the simple eddy viscosity formulation, instead incorporating the thermodynamics and physics of turbulent diffusion more accurately. From a modelling point of view, the implications of this are an increase in the number of equations there are to solve and a lot of extra coding! The benefits are a better model. The eddy viscosity is still there, but instead of being assumed to take a convenient form remains undefined. New equations are introduced that ensure the conservation of turbulent kinetic energy $\frac{1}{2}q^2$ and the quantity $q^2 l$ where l is a turbulence macroscale. Sometimes the dissipation rate of turbulence is also introduced. As these two new equations are introduced, so new diffusion coefficients are also introduced. The eddy viscosities and these two new diffusion coefficients are related to q and l typically by relationships derived from hypotheses necessary to close the system. These sophisticated models have only been possible to implement because of the very rapid advance in computing power and speed. Recently, newer relationships purporting to take stratification into account have been proposed. At the moment these are empirical and the contradiction of stratification and turbulence (one must physically destroy the other) in modelling terms is not resolved.

3.5.3 Modelling Fronts

Fronts are an important phenomenon to understand. They form at an interface between two water bodies that have different characteristics. At a front, there will be abrupt changes in density (salinity and temperature) and also velocity. There may also be significant upwelling or downwelling. A discussion of them in terms of physical oceanography could have taken place in the last chapter, but they are introduced here for reasons that will become apparent shortly.

Unless an analytical model is available, the modeller is faced with trying to formulate an accurate numerical model. Finite difference models that have enough resolution to identify the front as it moves and evolves are quite a challenge to the numerical analyst. The problem with most finite difference methods is that they inject quantities of numerical diffusion, and numerical diffusion is precisely what is least desirable in a model of a front. Fronts that diffuse are quickly fronts no longer, yet fronts are maintained in the real sea by the local physics. Special schemes have been developed and this case study shows recent developments based on the clear account of James (1996). The kind of scheme that is used is a forward time step but centred space step scheme incorporating upwind differencing. All of these terms have already been used, but to make things clear, consider the simple equation

$$\frac{\partial c}{\partial t} = -\frac{\partial}{\partial x}(Uc)$$

which is a one dimensional advection equation written in flux form (the flux form is met again in Chapter 5), U is taken as constant. The finite difference form of this would be

$$c_i^{s+1} = c_i^s - U\frac{c_i^s + c_{i+1}^s}{2} - |U|\frac{c_i^s - c_{i+1}^s}{2}.$$

However, although the CFL condition and stability criteria are satisfied, sharp gradients are not satisfactorily preserved. One answer is to use a TVD (Total Variation Diminishing) scheme. It is quite technical, but is given here for completeness. For a fuller discussion see James (1996). The particular Lax Wendroff scheme used is defined by:

$$(Uc)_{i+1}^{(1)} = \frac{U}{2}(c_i^s + c_{i+1}^s) + \frac{U^2\Delta t}{2\Delta x}(c_i^s - c_{i+1}^s).$$

This is used together with an upwind differencing scheme:

$$(Uc)_{i+1}^{(2)} = \frac{U}{2}(c_i^s + c_{i+1}^s) + \frac{|U|}{2}(c_i^s - c_{i+1}^s)$$

which in fact is another way of writing

$$\text{when} \quad U > 0, \quad (Uc)_{i+1}^{(2)} = Uc_i^n$$
$$\text{when} \quad U > 0, \quad (Uc)_{i+1}^{(2)} = Uc_{i+1}^n.$$

The TVD scheme is to allow

$$(Uc)_{i+\frac{1}{2}} = (Uc)^2_{i+\frac{1}{2}} + L(\alpha)[(Uc)^1_{i+\frac{1}{2}} - (Uc)^2_{i+\frac{1}{2}}]$$

where L is a limiter function which depends on the quantity α where

$$\alpha = \frac{(Uc)^1_{i+\frac{1}{2}-n} - (Uc)^2_{i+\frac{1}{2}-n}}{(Uc)^1_{i+\frac{1}{2}} - (Uc)^2_{i+\frac{1}{2}}}.$$

The subscript (or superscript) s is absent from both the Lax Wendroff and upwind schemes simply because all variables are at the current time step. The n on the right hand side is either $+1$ or -1 and is defined by

$$n = \text{sign}(U).$$

Several different functions $L(\alpha)$ have been tried, with varying degrees of success. One choice is

$$L(\alpha) = \frac{\alpha + |\alpha|}{1 + \alpha},$$

others are listed in James (1996). These kind of schemes are successful in keeping fronts sharp. It is impractical to apply such a complex scheme to a very large area, and indeed there is little evidence that success would follow. The way forward is to apply such TVD schemes to those finite difference points around the edges of the front so that the sharpness is preserved. Figure 3.6 shows how such a scheme allows a sharp front to propagate. There are many front preserving schemes, none is perfect and some perform better including a scheme based upon not linear but parabolic (quadratic) interpolation outlined by James (1996).

Improvements in finite difference schemes are occurring all the time and it is not long, given the fast communication today before ocean modellers are trying them out. Modelling fronts remains a severe test for the best modern numerical schemes.

3.5.4   Finite Element Methods and Applications to Tides

The details, such as they are, of this model are taken from several papers by the US researchers Johannes Westerink and Rick Luettich, notably Westerink *et al.* (1994). That paper contains a graded triangular mesh of staggering complexity. It is reproduced as Figure 3.7. The model is two-dimensional, but uses spherical co-ordinates that cling precisely to the curved surface of the Earth. The non-linear terms are included and there is a linear law for sea bed friction. The natural co-ordinates chosen are longitude and latitude, and although these co-ordinates, being curvilinear, give rise to extra unfamiliar terms in the governing equations, conceptually the idea is simple. The two-dimensional slab of ocean is divided into triangles. All variables are assumed to be independent of depth, so that all that needs concern us is how to approximate those variables associated

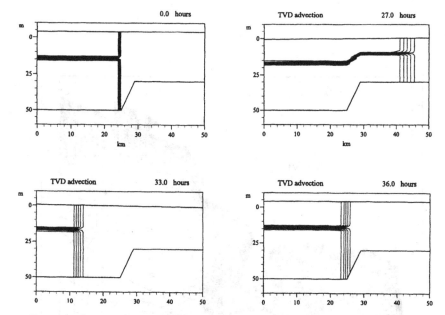

Figure 3.6: The sharp front modelled in four steps from 0 to 36 hours

with tides over each triangle (the question of boundary conditions is left until Chapter 4). The triangles are smallest, down to $5'$ arc length in those parts of the region where the variations are the largest, and conversely they are largest, resolution $1.6°$ where there is very little variation. There is a reason for this! In regions of high variability, there is a lot of structure in terms of changing currents and sea surface elevation. In order to capture this variation, many small triangles are required. On the other hand, in regions where variables are more or less constant, one large triangle that in effect connects the more interesting regions together will suffice. For tides, it is suspected that the shallower regions near the coast and close to islands are the regions of high variability. Within each triangle, therefore, variables such as the horizontal tide velocities and the tidal elevation (with respect to the geoid; that is, the equivalent to the standard tidal elevation except that a spherical coordinate system is being used) can take a particularly simple form. Westerink et al. (1994) use simple trigonometric functions; in papers that utilise cartesian co-ordinates, linear functions are popular. The important criterion is to ensure that all variables remain continuous at the borders of each triangle. For Westerink and his colleagues, this meant the elevation and velocities associated with the tide. This is consistent with the preoccupation of the paper, which was an accurate prediction of the tidal heights in sensitive areas such as around the land masses, particularly islands.

If it is required to predict higher-order quantities such as tidally induced residual flow, then it is essential for the variables in each triangle to be sophisticated enough for these quantities themselves to be continuous at the edges of the triangles. This means that the functions that represent each variables

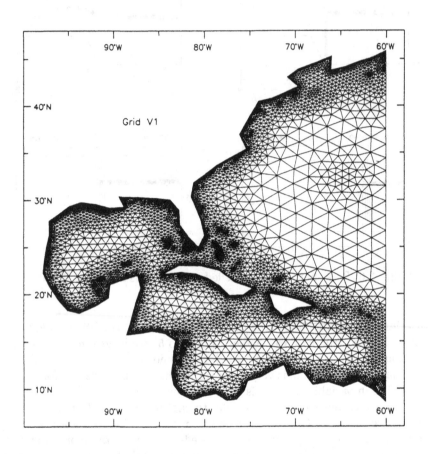

Figure 3.7: An unstructured finite element grid. From Westerink *et al.* (1994). Reproduced with permission.

in any particular triangle need to be more complicated. It is sometimes the case that shapes other than triangles are used in finite element simulations. The rectangle is quite a popular choice, when comparison with finite differences can be made easier, and near the coast, coordinates that fit the coastline more closely (termed isoparametric coordinates) can be very accurate and worth the extra effort. Finite difference schemes have been developed that use an irregular grid. The stability and truncation errors associated with such schemes are slightly more difficult to calculate, and keeping track in terms of numbering the points etc. can be tricky. Some regular shapes that do not consist of straight lines can be transformed into rectangles by using standard transformations. The transformed grids will be rectangular and amenable to standard finite difference techniques. This type of transformation technique works best however with man made regular shapes such as cylinders and spheres and not with the irregular natural shapes with which oceanographers have to contend.

It remains the case that finite differences are more popular than finite elements, mainly because of the bookkeeping associated with keeping track of the numbering of elements in regions where the shape is complex, but also because non-linear terms in finite element schemes lead to extremely high demands on computing power compared with finite difference schemes of equivalent complexity. However, see the next section: modern developments are changing this landscape.

3.6 Modern Software

One of the most used models is the Princeton Ocean Model (abbreviated POM) which is freely available on the web. The home page is

"http://www.aos.princeton.edu/WWWPUBLIC/htdocs.pom/".

A word of advice here, websites go out of commission from time to time so in the lapse between me writing this and you reading it things could have changed. The Princeton homepage "http://www.princeton.edu" is a safe start for a search if the POM is elusive. This provides the "front door" to the University from which hot links will be available. For information, the properties of the POM are now briefly summarised. It is three dimensional, with σ co-ordinates in the vertical. The non-linear terms are included in full. The full equations as are in Blumberg and Mellor (1987) so they will not all be repeated here. Included in it are extra terms F_x and F_y which are yet to be resolved horizontal diffusion processes. Temperature and salinity terms are included in the conservation equations. As these have not so far featured they are stated here. Some may think that these belong in Chapter 2, and they may be right. However, here they are: for temperature

$$\frac{\partial \theta}{\partial t} + u \frac{\partial \theta}{\partial x} + v \frac{\partial \theta}{\partial y} + w \frac{\partial \theta}{\partial z} = \frac{\partial}{\partial z} \left(K_H \frac{\partial \theta}{\partial z} \right) + F_\theta.$$

For salinity,

$$\frac{\partial S}{\partial t} + u\frac{\partial S}{\partial x} + v\frac{\partial S}{\partial y} + w\frac{\partial S}{\partial z} = \frac{\partial}{\partial z}\left(K_H\frac{\partial S}{\partial z}\right) + F_S,$$

where once again the unresolved horizontal diffusion processes are represented by the terms F_θ and F_S. The kind of expressions proposed for F_x, F_y, F_θ and F_S involve horizontal diffusion coefficients, although the beauty of the way the POM is written is that one is free to propose one's own model of diffusion. Turbulence closure is provided through a TKE (turbulence kinetic energy) scheme (see Chapter 4).

In the last few years the whole area of computational fluid dynamics has undergone a change. The power of available computing, particularly the UNIX based machine far exceeds that which could only be found on large mainframes less than ten years ago. The power of the PC is increasing at an even faster rate. There are now very powerful CFD (Computational Fluid Dynamics) packages that can be purchased and can be adapted to a limited extent to cater for your chosen problem. Two such CFD packages are marketed under the trade names of FLUENT ©, and CFX ©. A typical layout is that there is a central core programme that solves "standard" problems, then there are add-ons or extra modules that can be bolted on which are more specialist. Quite often the core programme has been developed for a specific purpose, for example modelling air flow in a restricted area which might be required for health and safety in the event of a fire. Such software is likely to be very sophisticated due to the resources available to produce it. There is no price too high to pay for safety! Advantage can be taken of this by adapting it or adding on bits of code so that it can be used for a wider class of problems. A modular approach gives the user more control over which bits to use and which bits not to use in a particular model. For example, there may be an entire module devoted to modelling the movement of sediment. This is crucial to engineers who wish to know about dredging and be able to assess how often it needs to be done, but is of no interest to those interested only in flooding. The quality of the graphics and the improved documentation is another very welcome development. It is still true however that the sellers of such software will not voluntarily point out shortcomings and tell the researcher under what circumstances the model does not work! He will not sell many boxes that way. The development of hybrid models (mixed finite difference/finite element for example) is still in its infancy, but is bound to come.

Chapter 4

Boundary Conditions

4.1 Introduction

We have been made aware during the last chapter of the importance of boundary conditions. The laws of motion attributed to Sir Isaac Newton have for 300 years quantified the motion of physical objects. The sea is a fluid which is also governed by these laws of mechanics. The basic equations, outlined in the last chapter are these laws together with the conservation of mass. The mathematical theory that underlies the motion of a frictionless fluid is called potential theory. Although the equations that govern the movement of the sea contain non-linear terms (namely $(\mathbf{u}.\nabla)\mathbf{u}$) if these are small and they usually are then potential theory tells us that the motion is determined by what happens at the boundaries. The way numerical schemes work confirms this (see section 3.3 on tidal modelling for example). Therefore the effects that are literally on the edge of models are in fact central drivers and play a major role in determining the resulting motion of the sea.

In this chapter we spend some time considering how to model the processes that occur at the boundaries of the sea. First let us distinguish between the different types of boundary. The sea-bed and coastlines are of course solid to fluid whilst the sea surface is fluid to fluid. However, modelling the physics of sea-bed as opposed to coastal boundaries is often very different. Modelling what happens at the sea surface could be the subject of a separate book or several books! It is one of the most complicated of boundaries and one is led to make drastic simplifications. However these simplifications are fully justified in terms of the specific phenomenon being examined. There are boundary conditions that have to be imposed because it is not possible to consider the entire sea; so-called open boundary conditions. There is also a boundary condition in time, the start condition. All of these demand separate consideration.

So, although the layman may believe that what happens on the boundary may not appear to be important, it actually drives the motion. It is all the more important, therefore, to understand precisely what is happening at the edges of regions. First, let us look at each boundary and distinguish a little more carefully between various types. Perhaps the most obvious boundary is the sea surface. It is also as we have said unfortunately, one of the most complex. Another obvious boundary is the coast. The coast is a solid boundary, and the treatment of solid boundaries is certainly less controversial. Another solid boundary is the sea bed; but we need to be careful to know what we mean by solid. Sea bed boundary conditions have occupied the attention of modellers for a very long time, and some quite sophisticated models are now in common use. The reason for this close attention is not hard to fathom; it lies in the role the sea bed plays in providing a sink for momentum. There is far more contact between the sea and the bed than the sea and its coastline, and a great deal of effort has been devoted to engineering problems associated with the sea bed such as dredging, erosion and scour. Conditions imposed at the sea surface, the coast and the bed are called closed because there is a definite physical edge that dictates to some extent the type of equation valid at the boundary (relating the surface current to the wind perhaps, or imposing no flow through a coast). There are also open boundary conditions. These are not actual, physical boundary conditions, but arise because domains that are not closed lakes or the entire globe have to have edges that cross the open sea. Open boundary conditions are necessary to apply to the open edges of models of these regions.

Another very different boundary condition arises form having time as a variable. All models have to start, and the state of the variables at time zero, the initial condition of a model is of course a boundary condition. In complex models that have significant non-linear terms, the start conditions are not important because the sea soon 'forgets' how it started to move and becomes, so to speak, wrapped up in its own dynamics. Systems or models with short memories are the type that lead to chaotic behaviour which is deterministic (that is, it lacks a random or stochastic element), but exhibits behaviour on many very different length scales. This is a popular topic nowadays, but the modelling of strongly non-linear systems is beyond the scope of this modest text. Having said this, in models that are not strongly non-linear, and that includes most of them, the start condition is very important. As examples, one can think of initial conditions that drive storm surge forecasts, diffusion models (see Chapter 5), and indeed weather forecasting itself, which would simply be a non-starter without accurate initial conditions.

The most successful models begin by being simple and idealised so that the essential dynamics is present in glorious isolation, uncluttered by awkward boundaries or complex but unimportant effects. The rectangular ocean models of yesteryear (see Chapter 6) provide particularly vivid examples of this. In very idealised models, the sides and bottom of the box-shaped ocean are flat, as is the surface. A rigid lid approximation is imposed whereby the sea surface is assumed solid. The open boundaries are simplified to be lines of flow (streamlines), and time does not feature. In spite of these idealisations, important features pertinent to the understanding of the general circulation of the ocean can still be predicted. Moreover, because of the elementary nature of the model, the user can see precisely what causes a phenomenon such as the western intensification of ocean currents.

As models are made more realistic, boundary conditions need to be treated with greater care. In particular, the different treatments of horizontal and vertical boundaries which stem from their differing scales are very important to maintain in ocean models. In models of less horizontal extent, such as an estuarial model, the distinction between horizontal and vertical boundaries become less important. In an estuary, there are in fact only 'water-air' and 'water-solid' boundaries. Another recent development in modelling which fits naturally into discussion of boundary conditions is data assimilation. Data assimilation is the incorporation of data into a model *as it runs*. It finds a natural home in limited area ocean models where a neighbouring model, or perhaps a larger one, can input information into the given model as it is running by having both models running in parallel. A second possibility is to input observational data into a model as it is running. For example, if a limited area ocean model is running, then as it evolves it could be possible for an eddy to migrate into the model even when the nature of the model makes the formation of such an eddy within the model impossible. This would be done by using as boundary conditions the velocity and elevation appropriate to an eddy along one of the model's edges.

Another obvious use of data assimilation is in storm surge modelling (see section 6.8). In storm surge models, it is wise to update the weather input as the model is running in order for the enhanced elevation to be predicted with the greatest possible accuracy. The increasing availability of satellite data make the improvement of model prediction by data assimilation a real possibility in areas where it has not been in use so far. However, for this text, data assimilation takes a relatively minor role as its main use is in obtaining answers that fit observations, and not answers that improve the understanding of the fundamental processes. Let us now return to the main subject of this chapter and examine the different types of boundary condition more closely.

4.2 The Sea Surface

The surface of the sea can be a flat calm, it can be mountainous, or any condition between the two. In ocean models it is commonly taken as flat, because over reasonably short periods of time, say a few minutes, the up and down movement averages to zero. Since ocean models are concerned primarily with bulk movements through ocean currents, time steps longer that this are used which implicitly implies averaging on the right time scale to eliminate the vertical movement of the sea surface. It is in Chapter 6 that simple models of the observed surface waves are introduced. These models are necessary for small scale motions such as wind rows (Langmuir circulation) and some estuarial dynamics. They must also feature in models that help the designer of offshore engineering structures, but they will not feature heavily in an oceanographic text. There is, of course, an entirely different vertical displacement of sea surface due to astronomical forces called tides. These need special attention and also feature in Chapter 6.

The surface of the ocean can thus be assumed flat (apart from tides). It can still move, however; it moves horizontally like an airport travellator. To force the ocean surface to act as if a solid barrier were against it is nowadays almost always unacceptable. This was called the rigid lid approximation and flourished briefly in the 1950's and 1960's. It is much more usual to allow the wind to act on the sea and to move its surface tangentially through frictional forces. The momentum thus introduced into the surface is then transferred to the ocean underneath by the turbulence so created. A way to achieve this is to relate the velocity at the sea surface to the sea surface stress via a law which might be a stress rate of strain relationship (Newtonian eddy viscosity), or to use a more sophisticated law. The simple but successful Ekman model has already been outlined in Chapter 2, section 2.6.1. The more complex laws at the sea bed are dealt with in the next section, and overall models that deal with coastal sea problems that require sea surface boundary conditions form part of Chapter 6.

It is also possible to allow for other boundary influences through the sea surface. If the temperature variation of the ocean is being modelled, the surface can be allowed to heat or cool due to outside influences (night and day, or the different seasons) and this can be incorporated by imposing temperature or temperature gradient conditions at the sea surface itself. A model would incorporate this via a source (or sink) term in the equation governing the diffusion of temperature downward through the water column. Other sea surface effects such as rainfall and evaporation have not been the concern of the modeller and will not be considered here, however they can be incorporated in principle using sea surface source or sink terms. In the newest ocean circulation models, the potential energy that arises from the rise and fall of the sea surface has an input into the salinity distribution. This is interesting in that models of the ocean circulation in the 1930's were entirely driven by the effects of precipitation, evaporation and freshwater inflow but these models were abandoned when the later wind-driven models seemed to be so much better at predicting the observed circulation. These latest models show us that we are wrong to reject completely such models. Instead, we must incorporate these effects alongside the more dominant wind driving in a more complete description of the surface (and coastal) boundary conditions.

4.3 The Sea Bed

The most obvious characteristic of the sea bed is that it is a solid barrier and that water must not be allowed to pass through it. This might seem obvious, but the smaller the scale of the model, the trickier such a criterion is to apply. A sea bed may first of all be steep, so that care needs to be taken that the boundary condition mentioned above involves the flow perpendicular to the sea bed. If the bed is flat, it may be sandy or muddy, in which case can it be assumed solid? Also, what about the cohesive properties (stickiness)? Mud is in reality a viscoelastic fluid with very complex properties but this elasticity will be ignored here. Viscosity is by far the simplest way that friction can be modelled. In a viscous fluid, all components of the velocity including those parallel to the sea bed, must be zero at the bed itself. Although the ocean is not a viscous fluid in the accepted sense, its equations are similar enough for such a boundary condition to hold. Finally, if very detailed modelling is to be considered, then some techniques from mechanical and aeronautical engineering modelling can be used. In particular, the defining of a roughness length to represent the character of the bed (sand, gravel or rocks), a laminar sub-layer where the flow regime is viscous but not turbulent, then a transition to turbulence when one is clear of the bed.

A quite recent successful modelling approach to the sea bed boundary has been to use $k - \varepsilon$ turbulence closure schemes (TKE schemes so called). Although the models that utilise these schemes apply to the entire sea, it is the fact that this particularly accurate model is required to simulate turbulence in areas that are dominated by frictional effects which causes us to go into more detail here rather than in Chapter 6, which might be thought the more natural home for them. The TKE (turbulent kinetic energy) models themselves were first developed by modellers of aeronautical systems where the ability to predict the detailed flow near critical parts of aeroplane wings was an essential requirement. In Chapter 2, we encountered the concept of eddy viscosity as a quantity that represented how stress is related to the rate of strain (or shear) of a turbulent flow. It is analogous to kinematic viscosity in laboratory viscous laminar flow, the principal differences being that turbulent viscosities are much larger than their laminar counterparts because turbulence is far more efficient at transferring momentum from one streamline to a parallel streamline. Also a turbulent eddy viscosity is not a fixed property of a fluid. (In laminar flow, the viscosity is as a fixed a property of the fluid as is, for example density. In turbulent flow, eddy viscosity can vary from place to place and can change with time.) Eddy viscosities cannot, unfortunately, account for the observed behaviour of a flow adjacent to the sea bed. Instead, modellers of recent times have used the above-mentioned, more sophisticated TKE turbulence closure models. What follows is a brief description of this more complex model of turbulence. By the very nature of these turbulence models, they involve complicated concepts. The mathematics is largely absent from the account here although there is a little more where the numerics are discussed in section 4.6. There is a fuller account in Blumberg and Mellor (1987) and Mellor and Yamada (1974).

The rationale behind most turbulence closure schemes is to start with the well known assumption of eddy viscosity, but then to introduce other variables that can be related to length and velocity scales which represent scales of turbulence. To be specific, if K_q represents the eddy viscosity (or eddy diffusivity), then:

$$K_q = lqS_a,$$

where L and q are appropriate length and velocity scales, respectively, and S_a is a factor that depends on the stability of the flow. This stability will, in turn, involve the density change with the vertical as well as the shear of the flow, $|\partial \mathbf{u}/\partial z|$. The velocity scale q and length l will themselves obey equations. For example, $\frac{1}{2}q^2$ represents the energy associated with the turbulence. In simpler models, S_a is simply a number and the complex structure of the turbulence model is carried in the two scales through the equations obeyed by them. It is quite usual these days for ostensibly simple two-dimensional models to have quite complicated turbulence closure schemes attached to them which can mimic successfully the momentum transfer just above the sea bed. This kind of model is often called '$2\frac{1}{2}$ – dimensional', and has the additional merit of being much cheaper to run (and in some cases more reliable) than more complex fully three-dimensional models. One interesting feature of virtually all sea bed models that purport to be detailed models of dynamics is the presence of a layer adjacent to the bed where the velocity profile is logarithmic. That is, the speed U is related to the distance from the sea bed z by an expression such as:

$$U = \left(\frac{\tau}{\kappa u_0}\right) \ln\left(\frac{z}{z_0}\right),$$

where τ is the sea bed stress, κ is a constant attributed to von Karman and usually given the value 0.4, and u_0 is a constant representative of the speed of the flow just above the roughness elements. These roughness elements are the sand, rock etc., that lie on the sea bed and interrupt the flow at the bed itself. The constant z_0 is a length that represents the average magnitude of the height of these roughness elements above the sea bed. At a height of z_0, the flow is no longer interrupted by sea bed debris. These ideas are displayed in Figure 4.1.

In addition to the correct modelling of the physics at the sea bed, at the bed itself some approximations are usually required because numerical methods which incorporate grid boxes (see Chapter 3) are being employed. This means that one point will be above the sea bed, but the adjacent point must be in the bed itself. For this reason a *slip velocity*, whereby the velocity at the sea bed is not zero but simply some value, is a common proposal. The value chosen is consistent with an appropriate quadratic or linear friction law that relates frictional stress directly to velocity at the bed. The usual controversy over whether to use such a law or whether to use a no-slip law at the sea bed itself is precisely equivalent to the imposition of a linear friction law a little above the bed at the nearest grid point. Having said all this, however, sophisticated turbulence closure schemes are now central to modelling the dissipation of momentum at the sea bed correctly, and numerical approximations of the sea bed boundary conditions merely alter the actual points at which these conditions are applied, not their application.

4.4   Coastlines

In the ocean, the coast and the edge of the continental shelf can be taken as being virtually synonymous. This is because of the magnitude of the horizontal length scales. If a model is to cover the entire horizontal extent of an ocean, perhaps 10000 km, with 100 points, then each point has to be 100 km apart.

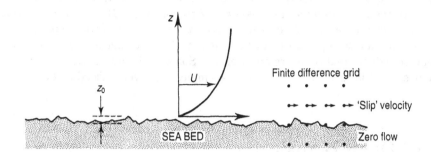

Figure 4.1: The sea bed, showing roughness length (z_0), finite difference grid and slip velocity.

This then means that a typical continental shelf is entirely lost between the coast and the first grid point. Under these circumstances the coast itself, can be assumed to be a shear vertical cliff. Exactly similar representations of the slip or no-slip boundary conditions at the sea bed are possible at the coast too when it is a vertical cliff. However, there is a subtlety that arises because in some models the Coriolis parameter varies with latitude which makes a no-slip condition dynamically distinct from a formulation which includes a slip velocity. However in a text on coastal modelling this does not feature strongly. There is no need of course for the complicated turbulence closure schemes in these large-scale models.

For smaller-scale models, the continental shelf must be taken into account. Normally models do not straddle the continental slope; they either lie entirely on it or it provides the location for boundary conditions of the deep sea models. Models of the continental shelf itself have open boundary conditions which need careful handling and deserve a section of their own.

At smaller scales still, say coastal inlets or estuaries, the coasts are treated in various ways. Often they are steep sided and a vertical face is allowable; on other occasions there may be mud flats or sandy beaches of minimal slope which demand that the depth of water itself is zero. There is a potentially serious problem if the sea dries inside a model region, because certainly in two dimensional models the depth occurs as a parameter in the denominator of various terms of the governing equations. One approach is to allow the sea to dry in cells in such a way as to compute the velocity as zero before (potentially) dividing by the zero depth. Research is still active, however, on the effects this has on the velocities elsewhere in the model domain. When there are drying regions present in a model, the horizontal domain of the problem changes with time (perhaps due to tides). Such a formulation is difficult to solve since the shape of the coast becomes an unknown of the problem. Dealing with moving coastlines can be technically difficult, but as a concept it (forgive the pun) holds water. In Chapter 6, more is said about models of continental shelf seas and what boundary conditions are appropriate. Some of the numerical applications are described later in section 4.6, and examples are given in section 4.8.

4.5   Open Boundaries

The boundary conditions considered so far can be considered *natural* in that they arise from a sea being surrounded by solid or air. It is the physics of the interface that by and large dictates the form that the boundary conditions take. Open boundary conditions occur simply because a model must end where there is no coast. Typical open boundaries are the edge of the continental shelf; Figure 4.2 shows a typical domain off northwest Europe. In this example, there is an extensive open boundary which more or less marks the European continental shelf. Estuarial models and models of coastal inlets will have open boundaries with the sea. Even deep ocean models which span the Atlantic or Pacific Ocean from west to east often have equatorial and northern open boundaries. In this latter case, the open boundary is dealt with in the same way as a solid boundary in that it is assumed to be a streamline. (No flow passes through a streamline, although in fact all that is required is that there is no *net* flow; there can be some transfer of fluid, but it has to be the same in each direction.) For smaller models however, the open boundary is treated differently. The essential feature a modeller wishes to retain is a lack of sensitivity as to where precisely the boundary is actually placed. If this is the case, then some confidence can be placed on the solution. Open boundaries are often subject to 'radiation conditions'. This is a way of allowing the open boundary to 'let out' flow etc., from the domain without reflecting any energy or momentum back into the domain. Numerically, this is hard to achieve exactly, but the further a boundary is away from the region of interest, the better (more accurate) the results. Another way is to 'nest' models such that a fine grid model is embedded inside a coarser grid model. Much care needs to be taken with this, but it is a preferred solution to simply extending the grid until it is 'far enough away'. For no matter how far away an open boundary, its effect can still travel through the entire domain well before the currents have settled down. In fact the effects due to the boundary have to attenuate, and this implies the presence of some kind of friction which may not be consistent with the physics of the sea. In the next section, we examine in general the numerical representation of the different kinds of boundaries, including the troublesome open boundary.

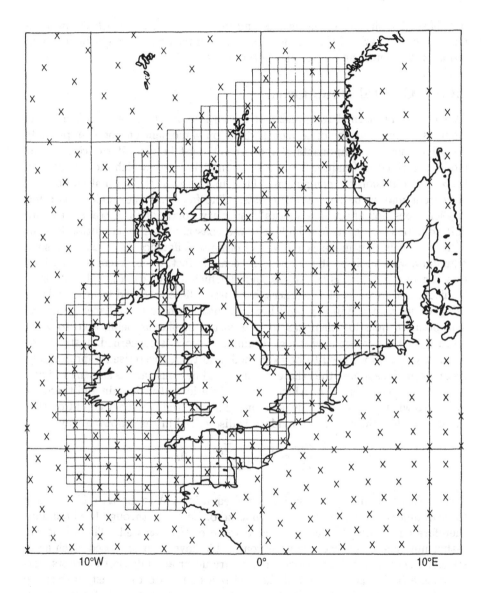

Figure 4.2: The continental shelf sea model (CSM) used as the basis for the IOS surge forecasts, with grid points (×) of the Meteorological Office 10-level weather prediction model which supplies the required forecast winds and atmospheric pressures. From Flather (1979), Reproduced with permission.

4.6 The Numerical Representation of Boundaries

In this section the preceding physics and modelling will be translated into numerical form. Some of the boundaries can be interpreted straightforwardly in terms of differences and let us tackle these first.

4.6.1 Coastal Boundaries

At a coast, the normal component of the current must be zero. If the model includes a representation of horizontal diffusion of momentum, for example eddy viscosity, then all components of the current must be zero at the coast. One model described in Chapter 3 was a semi-implicit model which did not contain horizontal diffusion of momentum. Thus in this case the normal component of current had to be zero at all coasts. The fact that the model was semi-implicit does not affect how coastal boundary conditions are translated into finite differences. Let us now look at how a coastal boundary condition is dealt with in models that make extensive use of finite difference schemes. At a solid boundary, whether it be the sea bed or a coast in fact, the problem is essentially that a grid point that is in the sea and at which an equation is valid needs to be discretised in terms of differences, but these differences need to be calculated using values that are in dry land because that is where the adjacent point is. The point in question then has to be rejected insofar as being able to write down discretised versions of the governing equations is concerned. Instead, the boundary conditions need to be examined, and to be discretised appropriately. As an example, suppose that there is a straight line boundary as shown in Figure 4.3. At the point marked A, which is in the interior of the domain and therefore in the sea where various equations are indeed valid, the boundary condition that there is no flow through the boundary is discretised in the following way:

$$\frac{\partial U}{\partial x} \approx \frac{U_{i+1} - U_{i-1}}{2\Delta x} = 0.$$

This replaces the full discretisation of the equations. Also, because U_{i-1} is due to be evaluated on dry land, it is taken as zero. Zero flow through the boundary therefore automatically leads to a zero value of U_{i+1} as well. This obviously rather unrealistic, and various ways round this have been proposed, such as to stagger the grid such that velocity points are never actually on the coasts, and to make a fictitious extension of the grid one point into the coast so that the domain is not artificially shrunk. The Arakawa C grid, outlined in Section 3.4.3, is one modern finite difference grid that facilitates this. Backhaus (1983) in his semi-implicit scheme used the Arakawa C grid, and at boundaries allowed for drying. The numerical scheme he employed was carefully designed such that as points "dried" no inconsistencies in terms of conserving transport of fluid occurred. Flather and Heaps (1975) give a good technical account of dealing with drying regions in a tidal model. In general, dealing with drying in a three

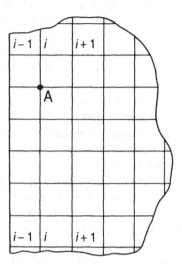

Figure 4.3: The finite grid near a boundary.

dimensional model is easier than dealing with drying in a two dimensional model as in the latter it is a true singularity whereas in the former it is not.

In problems that are primarily concerned with the modelling of waves, there is another way of dealing with geometrically simple boundaries, which in the context of this text means straight boundaries. If perfect reflection of the waves occurs at coasts, then it is permissible to allow the domain to extend as if the coast was a mirror with an image domain on the other side. The method of images is very well known in various fields of applied mathematics and can be applied in oceanography, but only limited to cases of simple geometry. Because of this, it is not extensively used.

4.6.2 Sea Bed Boundaries

At the sea bed, a similar problem can occur. The specification of the velocity just above the sea bed rather than actually on it is tantamount to specifying a slip velocity at the bed itself or at the nearest grid point above it, simply because the resolution of the finite difference scheme does not allow them to be distinguished. A neat trick mentioned briefly last chapter (see section 3.5.2) is to use something other than layers in the vertical, or at least to use a continuous representation of velocity close to the bed itself. One alternative which has proved very popular is to use a series expansion of variables in terms of functions of the vertical co-ordinate (or the sigma co-ordinate if this has been used).

If finite differences are used, then the sea bed boundary condition typically becomes the specification of stress at some point above the bed. Practically, however, this is not so much of a problem since our observational colleagues have to measure the stress at some point above the sea bed and not actually on it. The standard height above the bed for such measurements is 100 cm. It is then this measurement that is used as the boundary condition, and this is of course not controversial as it marries what is available to what is required by the modellers. The problem is that neither precisely matches reality, so that validation using, for example, a χ^2 test would not be valid. There is no null hypothesis! (See section 4.8). The sea bed boundary is therefore in mathematical terms treated in the same way as any coastal boundary. The essential difference has already been alluded to; the sea bed is a crucial momentum sink and a correct interpretation of the physics there is extremely important to ensure a good model. "Good models" for the physics of the sea bed date from the 1960s when TKE (turbulent kinetic energy) models were first developed to describe what was observed in laboratory models of fluid flow. In the early 1970s these were adopted for use first in the atmosphere and then the ocean. Researchers at the Princeton University's Geophysical Fluid Dynamics Laboratory in the USA then developed a hierarchy of models. Indeed a great deal of trouble has been taken in the last fifteen to twenty years to develop physically accurate but necessarily complex TKE schemes, so their numerical implementation has to be done carefully. It would be nonsense to throw away all the sophistication by representing them using ill-considered numerical approximations. As we have said straightforward difference scheme at the sea bed is precisely equivalent to imposing a slip velocity at the first wet grid point above the bed. Is there another way that this can be improved upon? One alternative which is simpler than refining the entire grid is to impose a quadratic friction law on the flow at this grid point, and demand that the C_D (drag coefficient) be related to a logarithmic law in the boundary layer under the grid point. This kind of modelling was done by Blumberg and Mellor (1987). For those interested in some details, these follow but can be skipped if desired. At the lower boundary the bottom stress (τ_{bx}, τ_{by}) is related to gradients in velocity through the standard law:

$$\rho_0 \nu_v \left(\frac{\partial u}{\partial z}, \frac{\partial v}{\partial z} \right) = (\tau_{bx}, \tau_{by}),$$

where ρ_0 is the ambient density. In addition the turbulent kinetic energy q^2 is related to the friction velocity $u_{\tau b}$ through the direct law

$$q^2 = B_1^{2/3} u_{\tau b}$$

where $B_1 = 16.6$ a value derived empirically. There is the fact the $q^2 l = 0$ where l is a turbulence macroscale and finally the vertical velocity w_b must be given by

$$w_b = -u_b \frac{\partial h}{\partial x} - v_b \frac{\partial h}{\partial y}$$

where $z = -h(x, y)$ is the equation of the bottom topography. Assume a quadratic friction law at the bed of the form

$$\tau_b = \rho_0 C_D |\mathbf{u}_b| \mathbf{u}_b$$

with the drag coefficient C_D given by

$$C_D = \left[\frac{1}{\kappa} \ln \left(\frac{h + z_b}{z_0} \right) \right]^{-2}$$

with the suffix b denoting numerical evaluation at the grid point nearest the bottom and κ is von Karman's constant. Then it is possible to derive the standard logarithmic layer at the bed in the form

$$\mathbf{u}_b = \frac{\tau_b}{\kappa u_{\tau b}} \ln(z/z_0)$$

provided there is enough resolution. Of course, such elaboration is over the top if the sea bed boundaries are not well resolved. In this case a numerical value of C_D say 0.0025 will do.

4.6.3 Open Boundaries

In this section we shall look at the open boundary. Essentially, when an artificial boundary straddles the sea it is the job of the fictitious condition that needs to be applied to mimic the motion of the (non-existent) sea on the outside. In particular, the condition must allow waves to pass cleanly out of the region without reflection. It must also cater for any incoming information, typically an incoming tide. Nevertheless the term "radiation condition" is the usual phrase used for this open boundary condition. Unfortunately in the situation where there are no incoming waves an ideal perfect radiation condition is only possible if the boundary is infinitely far from the source of any waves (see the monograph by Givoli (1992)).

From a purely practical point of view the two-dimensional equation obeyed at the open boundary is usually a simple advection equation of the form:

$$\frac{\partial u}{\partial t} + U \frac{\partial u}{\partial x} + V \frac{\partial u}{\partial y} = F,$$

where (U, V) is $(U_0 + c, V_0 + c)$, and (U_0, V_0) is a background flow, the local current near the boundary and c is the appropriate wave speed. The right hand side F represents incoming waves, invariably tides in a continental shelf model, and is obtained through observations or via running another model. The celerity c is $\sqrt{gh}$ for tides and similar water waves (see Chapter 6). In finite difference form, this condition might be written as:

$$U_{i,j}^{s+1} = U_{i,j}^s - \frac{U \Delta t}{2 \Delta x} \left(U_{i+1,j}^s - U_{i-1,j}^s \right) - \frac{V \Delta t}{2 \Delta y} \left(V_{i,j+1}^s - V_{i,j-1}^s \right) + F,$$

where we have used centred differences in space. Some kind of condition like this is essential to apply at open boundaries but perfection is not yet possible as hinted at earlier (Durran (1999)).

As a specific example, let us look at the TKE model of Blumberg and Mellor (1987). In this paper, two types of open boundary condition are prescribed, the inflow and the outflow. Temperature (T) and salinity (S) obey the equation

$$\frac{\partial}{\partial t}(T, S) + u_n \frac{\partial}{\partial n}(T, S) = 0,$$

where n denotes the normal to the open boundary. Turbulence kinetic energy $(\frac{1}{2}q^2)$ and the macroscale turbulence quantity $(q^2 l)$ are calculated with sufficient accuracy at the open boundaries. It is sufficient, even considering the sophisticated nature of the rest of their model to calculate theses two quantities neglecting the non-linear advective terms. Additionally a radiation condition of the type mentioned above is also imposed, viz.

$$\frac{\partial \eta}{\partial t} + (gh)^{1/2} \frac{\partial \eta}{\partial n} = F(s, t).$$

Intrinsic co-ordinates (s, n) are used where s is tangential and n is normal to the boundary. The forcing function $F(s, t)$ incorporates incoming tide together with any other currents. Once again, in their model it is permissible to neglect the non-linear advective terms in these open boundary conditions.

4.6.4 The Start Condition

The final boundary condition to consider is the start conditions. There is very little to say here, except that it is good practice to get as accurate data as possible to start a model. This is particularly true for storm surge models and ocean models that are not strongly non-linear. However, there is a class of models, (tidal models spring to mind) where the object of the exercise is to run the model in a predictive state to be as accurate as possible and *independent* of start conditions as soon after the start of the model as possible. For this kind of model, it is common to start the model by assuming a flat stationary sea and to allow the open boundary to input motion in order to *spin-up* the sea itself.

4.7 Finite element schemes

In finite difference schemes some details of which were given in the last chapter, the domain of the problem is divided into small areas (the elements), usually triangles, and simple functions are used to represent variables in each of the small areas. There are, of course internal borders between adjacent areas, but a good representation of reality is maintained by ensuring that at the junctions between the areas physical quantities such as fluid velocity, temperature and salinity are free of non-physical sudden jumps. At coasts however, things are different. The principle behind the finite element method is that, if the entire domain is considered - that is, the flow, temperature, etc. are each examined over the whole sea - then conditions can be imposed that correspond to basic physical laws such as the conservation of momentum and mass. This is done by forming a sum over all the elements and then applying each physical law. Boundary conditions such as no flow through coasts or a specified tidal amplitude have no place, directly, in such a scheme. In fact what happens is this. The equations that arise form the imposition of these laws over all the elements lead to enough equations to be able to determine every simple function in each small area. In fact, it turns out that there are *too many* equations because the numbering of the locations (nodes) where the variables need to be determined does not distinguish between which is an internal mode and which is actually on the boundary. This problem is overcome by writing down *all* equations (even those that are false equations) and replacing them by the appropriate boundary conditions written in terms of the function valid in these border elements.

Technically, the finite element representation of (say) tides is found by inverting a large matrix corresponding to solving a large number of simultaneous equations. Typically, there are 1000 equations with 1000 unknowns. Most of these equations contain only four or so unknowns, corresponding physically to any given unknown being dependent only on those variables in its immediate vicinity. The matrix is therefore largely full of zeros with non-zero entries clustered around the main diagonal. Mathematicians call these banded matrices, and there are special methods for dealing with them efficiently. The introduction of boundary conditions may lead to isolated off-diagonal entries which are potentially a nuisance, but not fatally so since judicious row operations usually restore the banded nature of the large matrix. These days there are in-built routines that generate finite element meshes that have inside them ways of incorporating different types of boundary conditions. Those interested in the technical detail should consult a specialist text such as Mitchell and Wait (1985).

The finite element scheme used by Westerink et al (1994) used highly unstructured graded grids based on a general wave continuity equation developed by Lynch and Gray (1979). The method is based on a clever combination of the basic equations which optimises the wave phase and amplitude characteristics. In general finite element methods are still subject to instability, but the scheme used by Westerink *et al.* (1994) is stable. The authors quote a Fourier analysis in constant depth using linear interpolation which indicates that a tidal wave is resolved with 25 nodes per wavelength. It is still necessary for the Courant number, here defined as $\sqrt{gh}\Delta t/\Delta x$ to be less than or equal to one. The stability analysis is similar to that for finite differences on a B grid.

For the oceanographic modeller, the most convenient aspect of finite element modelling is that the boundary conditions are embedded in the equations themselves and are not a separate feature, as is the case with finite differences. Of course there is the temptation to think of finite elements as in some way 'more natural' because of this, but this is a dangerous illusion, since both are as good or as bad as each other!

Validation studies (see the next section) have taken place for finite element models. For the study of Westerink *et al* (1994) outlined in the last chapter, the output of the model was compared directly with well known measured tidal constituents. In this model, a proportional standard deviation was defined as follows:

$$E = \left(\frac{\Sigma_{l=1}^{L} \left(\eta^c(x_l, y_l) - \eta^m(x_l, y_l) \right)^2}{\Sigma_{l=1}^{L} \left(\eta^m(x_l, y_l) \right)^2} \right)^{1/2},$$

where the symbols have the following meanings:

$$\begin{aligned}
L &= \text{total number of elevation stations within a given region;} \\
(x_l, y_l) &= \text{the location of the elevation station;} \\
\eta^c(x_l, y_l) &= \text{the computed elevation amplitude at a given station;} \\
\eta^m(x_l, y_l) &= \text{the measured elevation amplitude of a given station.}
\end{aligned}$$

In the model, a number of tidal constituents were simulated. In particular, the average error as measured by E was between 18.2% and 45.3%, where eight tidal constituents were considered and the entire domain was covered. Nothing statistical was attempted, but the regions of greatest error coincided with the amphidromic points (where the tide vanishes). Also, those stations with poor convergence properties (i.e. the model elevation only slowly converging to an answer) gave the poorest comparison with measurement.

For wind-driven flow, a statistical comparison between model output and observational data is hardly ever done since observations usually drive the model rendering them dependent and rendering statistical approaches invalid. The future availability of satellite data could herald the onset of proper validation studies. For the model of Westerink *et al* (1994), not only were validation tests in the form outlined above carried out, but sensitivity tests were also done. This took the form of splitting elements and examining the subsequent changes in both the amplitude and phase of the tide.

4.8 Model Validation

The validation of a model is the comparison of model output with what can be termed current knowledge. This knowledge usually comes from observation (see Chapter 1). It might seem as if nothing can be easier than to compare the output of a model with observations which usually occur in the form of data. However, this is not the case except perhaps for tides where there are very long and very accurate records (in most instances). If a model is run, and values of the variables are obtained, one is faced with the question of how accurate these results are. To take a specific example, a domain may be overlaid with a two-dimensional grid (Figure 4.3 is typical) and values of surface elevation, eastward and northward velocity obtained as model output at all grid points. How good are the results? Perhaps an observational programme over the same area of a convenient cruise has occurred and instruments have produced some data. First, the data will not have been taken at precisely the same locations as the model grid points; second there will in all likelihood not be enough data; and third not all of the data will have been collected at the same time. All of these factors make comparison between data and model output difficult. In order for some values to be compared directly, interpolation has to take place. Interpolation is a numerical technique that enables values of a variable at some intermediate location to be computed from those that surround it. It is a generalisation of curve fitting and there is now quite an extensive library of software that can help with this process (terms associated with interpolation include splines and least squares, both of which are useful, and Lagrange interpolation, which should be avoided). This then deals with one problem, but if there are not enough data, or they were recorded at the wrong time, there is not much we can do. Satellite obtained data promises to be very useful here and could provide a leap in the sophistication of model validation. Suppose that we have adequate data, what do we do about comparing these data with the output from the model? The easiest thing to do is to 'eyeball' both and assess the significance of any differences. This is still often the only method of validation used in marine science. While not excusing this very unscientific methodology, it is perhaps understandable that oceanographers and ecosystem modellers should be unwilling to expend large amounts of time and energy on sophisticated statistical techniques when much of the data are so roughly hewn from the sea. Nevertheless, there are techniques that can be of some use. If a direct comparison between model output and observations is possible, then one can analyse the differences between them statistically. For example, using the statistics of sampling, it is possible to tell whether differences are significant by using the t-distribution and to place confidence limits on the significance of these differences. If, further, it is possible to postulate that, as a *null hypothesis*, there should be agreement between observations and model output, then one can define this measure of agreement via the χ^2 test. This is a statistic that is defined by the expression:

$$\chi^2 = \sum \frac{(\text{Observed - Expected})^2}{\text{Expected}}.$$

The 'observed' values are the model results, and the 'expected' values are the corresponding observations. The $\sum$ sign denotes summation over all data points. Remember one technical point here: in order to use the χ^2 test the data has to be ranked, classified or otherwise rendered free of dimension. There are also questions of degrees of freedom. Consult Chapter 8 or statistics textbooks for further details. Of course, one could swap the role of model results and observations; this would in theory 'test' the observations against the model, assuming the model to be correct. If both are reliable, then there is no problem. If neither is reliable, then we get the standard arguments between those who measure and do field-work, and those who model and do calculations. Once the χ^2 statistic has been calculated, it is simple procedure to look at a table (see for example, Murdoch and Barnes (1974) *Statistical Tables*) and ascertain whether or not departures are significant. Of course, it is tempting to regard this as a definitive argument for or against a particular model. In reality, standard statistical tables have built into them certain assumptions involving the normal distribution that particular observations or model output may disobey. As a general rule, it is most unwise to use statistical methods that are more complicated than is warranted by the veracity of the data! If in doubt, contact a

professional statistician. A necessarily very brief introduction to some statistics
is given in Chapter 8.

Chapter 5

Modelling Diffusion

5.1 Introduction

One of the clearest indications that there is human habitation on this planet is the presence of artificially produced material in the world's oceans. For mankind, this is of course of central importance, also of more immediate concern is the presence of such material in coastal seas and estuaries. Sadly, much of this material is often poisonous to some degree as has already been mentioned in the first chapter of this book. The name *pollution* has been coined to describe foreign often toxic material in the sea. Once pollution is present, it does not remain unchanged but is pulled and pushed around by the currents and waves of the sea, spreading and (usually) diluting. This spreading takes place even if the sea were to be quiescent. The name of this process is *diffusion*. Molecular diffusion can be observed if a grain of potassium permanganate (purple) is placed in still water. A purple patch gradually grows. Of course this growth is enhanced if there are currents present. The school experiment that demonstrates convection by dropping a crystal of $KMnO_4$ (potassium permanganate) in a beaker of water being heated from beneath by a bunsen burner clearly shows enhanced spreading. Similarly in the environment pollution can be made to spread effectively by the action of strong and usually variable currents. A word must be said here about the use of the word *dispersion*. Quite rightly dispersion is used as a synonym for diffusion. However, dispersion has come to have a special meaning for physicists and applied mathematicians, it means the change (increase) in the wavelength of a wave or group of waves as it propagates (see Chapter 6). This spreading of the waves is dictated by the dispersion relation that gives the relation between wavelength, frequency and other relevant physical parameters. In view of this and subsequent possible confusion the word dispersion will not be used as meaning general diffusion.

5.2   The Process of Diffusion

Although diffusion can and does take place at the molecular level at sea in the same way as in the laboratory, far more important is the diffusion of pollution through turbulence. Turbulence, the random commotion of water, is ideal for diffusing pollution and it can do it reasonably successfully. It can act in a similar way to molecular diffusion but at far larger scales and the effect can be up to one thousand times greater. As you may suspect however, the story is by no means a simple random spreading but a combination of all kinds of different complex mechanisms. We shall be trying to model some of these mechanisms in this chapter. Other effects, not in themselves diffusive, can greatly enhance or in some cases inhibit diffusion. At a shear, which is often present near a boundary, the unidirectional current varies from zero on the boundary itself to a large value only a short distance from the boundary. A shear is particularly efficient at diffusing any pollution. It is far more efficient, for example, than turbulence, which is suppressed near boundaries (solid ones at least). On the other hand, there are convergence zones in some flows (for example Langmuir circulation; see Chapter 6). These convergence zones can act anti-diffusively, especially for buoyant contaminants. Their ability to re-concentrate hitherto dilute toxic substances is one of the main reasons for their study.

As far as modelling diffusion is concerned, one rather different species of model is the *particle tracking* model. Particle tracking or Lagrangian models are well suited to modelling the diffusion of particles because their *modus operandi* is to follow individual particles. Since the pollution can be simulated as a collection of marked fluid particles, these can be tracked by the model and various parameters (size of patch, location of its centre, etc.) output at appropriate stages.

To the fluid mechanist, diffusion is merely a consequence of applying the laws of fluid motion. Granted, they need to be applied with precision, so that effects mentioned above such as turbulence and current shear near boundaries are adequately represented, nevertheless diffusive effects should appear as consequences. To the practical modeller of real diffusion of real pollution this is not useful. In real life, too much is happening for everything to be included even in the most advanced of models. For example, an oil slick is one liquid, (oil) interacting with another (sea water) in a complicated way. Even the oil is probably composed of several varieties which over time separate into liquids which have a wide variety of properties, tar is very different from light crude oil. Then there is all the chemically induced flow caused by reactions, the heat generated by these reactions and various combinations of these two effects. Biology, especially the microbiology of bacteria and cells provide yet another possible source of flow. Accurate hydrodynamic modelling of this is still very far from possible.

However, the fluid mechanist's view is useful in one respect. That is, diffusion is a turbulent process that happens once a parameterisation of turbulence is included in the dynamic balance. The simplest model of diffusion is one in which diffusive transfer of material (through action) is directly related to the gradient of the concentration. The constant of proportionality governs how quickly the diffusion occurs, and is called the diffusion coefficient.

5.2.1 Fickian Diffusion

This kind of simple model can be used to predict the diffusion of outfall material due to sewage or industrial waste spillage in a river, provided the spillage is large enough, homogeneous enough, and provided the river flow is reasonably uniform. This model is termed Fickian (after A. Fick, who developed the idea in the mid-nineteenth century).

Let us derive from first principles the effects of turbulence on a contaminant (pollutant). A much simpler dimensionally based argument will be given a little later. Suppose a uni-directional flow can be split into a mean flow and a fluctuation due to turbulence. This was introduced in chapter 2. So let

$$u = \bar{u} + u'$$

where $\bar{u}$ is a mean and u' is a fluctuation about the mean ($\overline{u'} = 0$ of course). The pollutant has concentration c, and because this foreign matter is caught up in the local motion of the fluid, it too will have a mean and fluctuating part, so:

$$c = \bar{c} + c'.$$

Of course, nothing can be said about the relationship between c' and u', $\bar{c}$ and $\bar{u}$ without further assumptions. All we know is that $\overline{c'} = 0$ too by *a priori* assumption.

For simplicity let us assume that the flow is uniform across an area A. This is not essential, but by doing so we avoid ugly looking and potentially confusing multiple integrals. Thus the flux of contaminant across this area which is perpendicular to the direction of the current u will be Auc. The flux per unit area averaged over a suitable time is thus $\overline{uc}$. Some simple algebra reveals that:

$$\begin{aligned} \overline{uc} &= \overline{(\bar{u} + u')(\bar{c} + c')} \\ &= \bar{u}\bar{c} + \overline{u'c'}. \end{aligned}$$

The term $\bar{u}\bar{c}$ is called the *advection* of contaminant and might be what the intelligent layman would expect the flux of pollutant to be. The term $\overline{u'c'}$ is the *diffusion* of the contaminant and is due to the interaction between the two turbulent fluctuations. It is not surprising that such diffusion takes place, and for those with some knowledge of statistics a parallel with covariance of the two random variables can be drawn. Having introduced advection and diffusion through a simple one-dimensional model, we now move into three-dimensions. If a cube of sea (or estuary for that matter) is considered, then the total amount of pollutant entering it must be the same as the total amount of pollutant that leaves it. Calculus type arguments in a box of dimensions δx, δy, δz (see Figure 5.1 and Proudman (1953) chapter 6), give on taking appropriate limits:

$$\left(\frac{\partial}{\partial t} + u\frac{\partial}{\partial x} + v\frac{\partial}{\partial y} + w\frac{\partial}{\partial z}\right)c = 0.$$

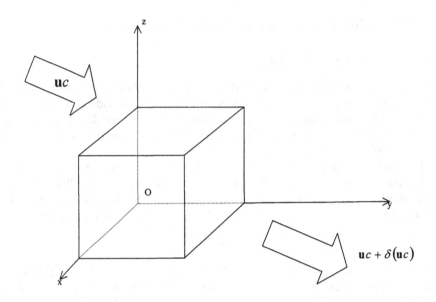

Figure 5.1: The Conservation of Pollutant

Now, writing $u = \bar{u} + u'$, $v = \bar{v} + v'$, $w = \bar{w} + w'$ together with $c = \bar{c} + c'$ and also noting that:

$$\frac{\partial u}{\partial x} + \frac{\partial v}{\partial y} + \frac{\partial w}{\partial z} = 0$$

shows that in order that actual flows into and out of the cuboid are the same we must have:

$$\left(\frac{\partial}{\partial t} + \bar{u}\frac{\partial}{\partial x} + \bar{v}\frac{\partial}{\partial y} + \bar{w}\frac{\partial}{\partial z}\right)\bar{c} + \frac{\partial}{\partial x}(u'c') + \frac{\partial}{\partial y}(v'c') + \frac{\partial}{\partial z}(w'c') = 0.$$

We now write:

$$\frac{D}{Dt} = \frac{\partial}{\partial t} + \bar{u}\frac{\partial}{\partial x} + \bar{v}\frac{\partial}{\partial y} + \bar{w}\frac{\partial}{\partial z}$$

which is the usual notation for the total derivative (also called differentiation following the motion or Lagrangian derivative). We also define:

$$u'c' = -\kappa_x\frac{\partial\bar{c}}{\partial x};\ v'c' = -\kappa_y\frac{\partial\bar{c}}{\partial y};\ w'c' = -\kappa_z\frac{\partial\bar{c}}{\partial z},$$

where κ_x, κ_y and κ_z are the diffusion coefficients. These three relationships are of the type originally proposed by Fick in 1855, but he was considering molecular diffusion. When dealing with turbulence induced diffusion it is necessary to define separate quantities for, as we have already seen in defining eddy viscosity which has been done in Chapter 2, turbulence induced diffusion is usually by no means independent of time or space. We have defined three of them in the three different directions. The vertical diffusion coefficient κ_z will always be distinct from the two horizontal coefficients κ_x and κ_y. In the open sea, there may be arguments to support putting $\kappa_x = \kappa_y$, but in an estuary for example the diffusion along the line of the estuary will be very different (much larger) from the diffusion across the estuary.

With these Fickian assumptions, the equation for $\bar{c}$ can be derived as:

$$\frac{D\bar{c}}{Dt} = \frac{\partial}{\partial x}\left(\kappa_x \frac{\partial \bar{c}}{\partial x}\right) + \frac{\partial}{\partial y}\left(\kappa_y \frac{\partial \bar{c}}{\partial y}\right) + \frac{\partial}{\partial z}\left(\kappa_z \frac{\partial \bar{c}}{\partial z}\right).$$

For some more detailed sensible discussion, see the textbook by Lewis (1997). With κ_x, κ_y and κ_z (unjustifiably) equal and constant, this reduces to what is known as the "advection-diffusion equation"

$$\frac{D\bar{c}}{Dt} = \kappa \nabla^2 \bar{c}.$$

Finally, if the advection terms are ignored, which may be justified for very slow flow, we obtain the standard diffusion equation:

$$\frac{\partial \bar{c}}{\partial t} = \kappa \nabla^2 \bar{c}.$$

We shall return to look at the solution of the advection-diffusion equations a little later. However, let us pause for some much simpler dimensional analysis.

If discussion is centred on the diffusion of momentum, the diffusion coefficient is our old friend eddy viscosity. The dimensions of the quantity, found in Chapter 2, are $L^2 T^{-1}$ (or perhaps density times this, $ML^{-1}T^{-1}$, which is the dynamic version). If, on the other hand, discussion focuses on a passive quantity, say temperature, salt or in fact any contaminant, then the amount of this passive quantity in the volume exhibited in Figure 5.1 which is considered free of sources and sinks must be conserved. In its simplest form, this means that the time rate of change of the concentration of this passive quantity at any particular point must balance the spatial gradient of, not the concentration itself, but the agent that causes the diffusion of the concentration. In an open ocean or sea, this agent is usually he turbulence-induced eddies. This agent is usually parameterised as being proportional to a constant times the concentration gradient (the Fickian assumption) as explained above. The balance is thus:

time rate of change of (concentration of passive quantity)
= gradient of [agent that causes the diffusion of the passive quantity].

In turn, the square bracket is a constant times the concentration gradient. If the passive quantity is labelled C, then in purely dimensional terms we have the dimensional version of the diffusion equation:

$$\frac{1}{T}C = \frac{1}{L}k\frac{C}{L},$$

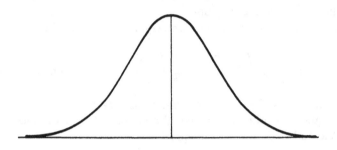

Figure 5.2: Gaussian profile

where k is the constant diffusion coefficient arising out of the Fickian assumption. Once again, therefore, it is seen that the dimensions of k are $L^2 T^{-1}$. This is now seen to be independent of the identity of the diffused quantity.

We have seen the simplifications that have to be made in order to derive the diffusion equation itself. The standard textbook for the solution of the diffusion equation in all its attendant forms and under all kinds of boundary conditions is Crank (1975). The often quoted solution to the one-dimensional diffusion equation in the form:

$$\frac{\partial c}{\partial t} = \kappa \frac{\partial^2 c}{\partial x^2},$$

(where the overbar has been dropped for convenience) is:

$$c = \frac{A}{\sqrt{t}} e^{-x^2/4\kappa t},$$

where A is an arbitrary constant. This solution describes the diffusion away from the source which is concentrated at $x = 0$. The diffusion is one-dimensional, hence is either a streak along the centre line of an estuary, or more realistically a cross stream average diffusion. The solution is not valid at time $t = 0$, except that it describes a fictitious infinitely strong source of infinitesimally small extent. (Those familiar with Dirac-δ functions may recognise an impulse when they see one.) At subsequent times however the (mean) concentration c has Gaussian profile, see Figure 5.2. Of course, this is exceptionally idealised. Even if the diffusion equation is valid, it is additionally necessary for idealised boundary conditions to hold. Nevertheless there are certain useful deductions one can make. Certain advanced mathematical methods can be used to deduce that, no matter what initial distribution of concentrate, the spread is still at the same rate. The Gaussian model is valid in one dimension, in cylindrical geometry in two dimensions and in spherical geometry in three dimensions. The analytical solution in two dimensions is involved, but the tree dimensional solution is quite straightforward. The equation is:

$$\frac{\partial c}{\partial t} = \kappa \left(\frac{\partial^2 c}{\partial r^2} + \frac{2}{r} \frac{\partial c}{\partial r} \right)$$

assuming spherical symmetry, with solution:

$$c = \frac{A}{r\sqrt{t}} e^{-r^2/4xt}.$$

In other words, there is simply an extra r in the denominator. Here we have written $r^2 = x^2 + y^2 + z^2$; r is the distance away from a point source at the origin. There is little direct marine application of such point sources in three dimensions. It is background material. Let us get back to the general advection-diffusion equation:

$$\frac{Dc}{Dt} = \frac{\partial}{\partial x} \left(\kappa_x \frac{\partial c}{\partial x} \right) + \frac{\partial}{\partial y} \left(\kappa_y \frac{\partial c}{\partial y} \right) + \frac{\partial}{\partial z} \left(\kappa_z \frac{\partial c}{\partial z} \right).$$

There is hardly ever any need for cylindrical or spherical geometry in marine science. One practical feature is that if pollution is spreading from an axisymmetric source in a river or estuary, and the coasts are far enough away not to have affected the behaviour of the contaminant, then the spreading does tend to be Gaussian. Assuming a variance σ^2 and a diffusion coefficient κ, it is possible to define κ by:

$$\kappa = \frac{1}{2} \frac{d}{dt} (\sigma^2)$$

from which on integration

$$\sigma^2 = 2\kappa t.$$

This is consistent with the theoretical (one-dimensional) Gaussian distribution cited earlier. If advection is added, but in the simplest possible manner then we arrive at an equation of the form:

$$\frac{\partial c}{\partial t} + U \frac{\partial c}{\partial x} = \kappa \frac{\partial^2 c}{\partial x^2},$$

where U is the speed of the stream. This stream is constant and in the direction of the spread of the pollutant. This simulates the spread of pollutant in a flowing river. The solution is:

$$c = \frac{A}{\sqrt{t}} e^{-(x-Ut)^2/4\kappa t}$$

and is not surprising. It represents a Gaussian spreading being transported (advected) downstream with speed U. More realistic models involve representing the all important shear present in most estuaries and coastal seas. As a mechanism for diffusing a pollutant which is not yet uniformly distributed entirely across an estuary or throughout the depth of a sea, shear tends to dominate where is exists. We return to it later in this chapter but meanwhile let us look at practical studies.

One of the most cited papers on diffusion was written by Akira Okubo in 1971. He, along with several other authors, recognised that oceanic diffusion occurs through many different mechanisms: turbulent eddies, shears of several origins, as well as biological agents. Obviously, it is not possible for a single model to simulate every diffusive effect. Okubo took on the task of examining many different measurements. In order to make some sense of all these data, it is useful (one could say essential) to provide a benchmark on which to measure and compare. Most experiments, certainly all those considered by Okubo, involve observing the spreading of a patch of dye. A patch of dye has a centre of mass, and a distribution of mass about this centre. All distributions have a variance which measures how spread out they are. As time progresses, the patch of dye increases in size, spreading out. It seems sensible to ask, therefore, what relationship there is between this measure of spreading and time. The straightforward Fickian diffusion, which is diffusion by turbulent eddies in the absence of shears, can be represented exactly by the diffusion equation as seen above and solved. The result of this solution is that variance is proportional to time (the precise definition of variance is given in Chapter 8). Okubo's results give that variance is proportional, not to time t, but to time raised to the power 2.3, $t^{2.3}$. The graph from Okubo's paper is given in Figure 5.3.

It can be seen from this graph that much of the data are scattered, and the line which corresponds to variance being proportional to $t^{2.3}$ is in reality a line of best fit or regression line. It will also be noticed that both axes are logarithmic which enable a power law to become a straight line. For those unfamiliar with statistical notions such as this, a summary can be found in Chapter 8. However, here we are more concerned with the mechanisms of diffusion and trying to model them. Figure 5.3 exhibits large scatter, even larger if one considers the logarithmic axes. However the same axes show a vast range on length and time scales over which the power law is approximately valid. The validity of this power law for small scales is very open to question. Okubo shows no data for scales of 10 m or less, even those that are shown are very scattered around the 50 m length scale. It is therefore reasonable to deduce that the diffusive mechanisms are most distinct at these very different length scales, and Okubo's power law, although very seductive, is too much of a simplification of reality. We look again at Okubo's model after the introduction of more theoretically based models. A simple Fickian law for example will always give variance proportional to t.

5.2.2   Shear Diffusion

Shear is the general name given to a change in the magnitude of a current perpendicular to the direction of flow. It is easier to see this than to describe it, so

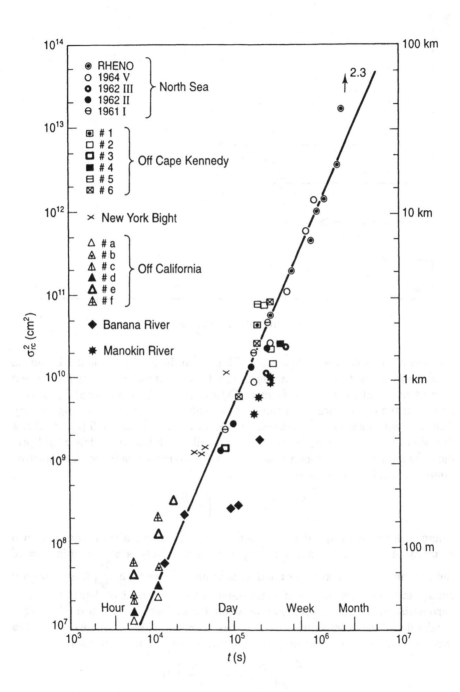

Figure 5.3: Diffusion diagram for variance versus diffusion time. From A. Okubo (1971). Reproduced with kind permission from Elsevier Sciences Ltd., The Boulevard, Langford Lane, Kidlington OX5 1GB, UK

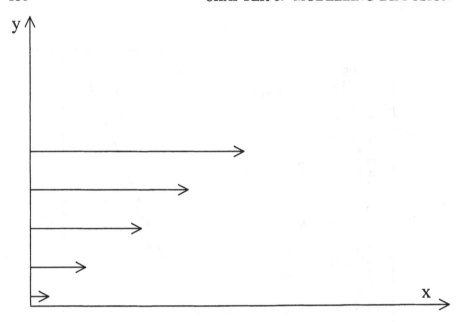

Figure 5.4: A shear flow.

a simple shear is drawn in Figure 5.4. That it can be a very efficient mechanism to mix and hence to enhance diffusion can be seen as follows. Consider a shear flow as depicted in idealised form in Figure 5.4. If this is caused by a wind acting on the sea surface, then there will be enhanced diffusion (see Figure 5.5). The diffusion takes place in stages, but the net result, Figure 5.5 (a) to (c) is a diffusion many times more rapid than most direct diffusion models would predict. To model such a process in a simple way is quite a challenge. The starting point is the advection-diffusion equation

$$\frac{\partial c}{\partial t} + u\frac{\partial c}{\partial x} = \frac{\partial}{\partial z}\left(\kappa_y \frac{\partial c}{\partial z}\right),$$

where u is the fluid speed in the x-direction this time taken as an unknown quantity. Any diffusion that is taking place horizontally is neglected in view of the presence of u which is assumed to dominate. If we take $\frac{\partial c}{\partial x}$, the horizontal change in the concentration of contaminant to independent of depth, then it is permissible to integrate vertically between sea surface and sea bed. The right hand side diffusive term integrates to zero provided there is no sea bed or sea surface input, which is certainly generally true. the result is therefore:

$$\frac{\partial <c>}{\partial t} + <u>\frac{\partial c}{\partial x} = 0$$

where $<c>$ and $<u>$ are depth mean values. If c and u are split into depth mean values, with the addition of variations from the mean, viz:

$$c = <c> + c'$$

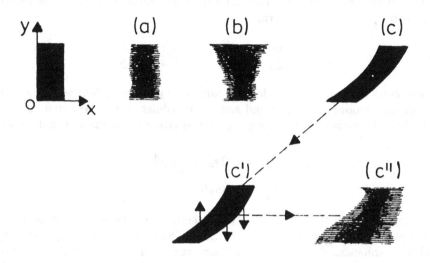

Figure 5.5: (a) Vertically distributed pollutant, (b) Pollutant sheared by a shear flow, (c) Pollutant diffused by vertical mixing.

$$u = <u> +u'$$

where u' and c' are quantities representing the departures of u and c respectively from their depth mean values. Analysis can be performed on these equations (see Lewis (1997)). In particular, in the spirit of Fick, an "effective" diffusion coefficient κ_e can be defined by:

$$\overline{u'c'} = -\kappa_e \frac{\partial c}{\partial x}$$

which encapsulates the interaction between vertical shear and the mixing of the contaminant. Lewis (1997) quotes a simple example for which

$$\kappa_e = \frac{<u>^2 h^2}{30\kappa_z}.$$

This is a simple linear shear ($2 <u>$ at the sea surface and zero at the bed) and κ_z is constant through the depth h. Although there are some limits in the above passages about a possible increase in the diffusion of a passive contaminant, nothing further has been said about the time a patch takes to diffuse. All models so far discussed lead to the variance of a patch increasing linearly with time. All that the above simple shear diffusion model gives is, under some circumstances, an enhanced diffusion coefficient. Suppose however, we assume a constant value for the vertical shear $\frac{\partial u}{\partial z} = \chi_z$ say. This is consistent with flow close to the sea bed. With constant diffusion coefficients too, the advection diffusion equation take the form

$$\frac{\partial c}{\partial t} - \chi_z \frac{\partial c}{\partial x} = \kappa_x \frac{\partial^2 c}{\partial x^2} + \kappa_y \frac{\partial^2 c}{\partial y^2} + \kappa_z \frac{\partial^2 c}{\partial z^2}.$$

Under certain assumptions to do with partitioning of energy between large scale eddies that produce the shear and small scale eddies that merely contribute to diffusion, the variances of the solution to this advection-diffusion equation are:

$$\sigma_x^2 = 2\kappa_x t + \frac{1}{6}\chi_z^2 t^3$$

$$\sigma_y^2 = 2\kappa_y t.$$

The first terms on the right are to be expected, but the term $\frac{1}{6}\chi_z^2 t^3$ is due to the shear in the x (longitudinal) direction. The t^3 dependence here certainly indicates enhanced diffusion. The diffusion coefficient κ_e (an effective diffusion coefficient due to shear) defined earlier can be calculated to be proportional to t^2 under some circumstances.

5.2.3   Homogeneous Diffusion

Let us go back a stage. Now that we have seen the kinds of models that are available, it is useful to return to a more fundamental approach and look again at the process of diffusion from first principles. It was Batchelor (1953) who first brought the ideas of the notable Russian mathematician and statistician A N Kolmogorov (1903 - 1987) to the general public. In particular his work first published in 1941. The central idea is that in a field of turbulence locally all is the same, independent of the direction considered. The name homogeneous or isotropic turbulence was coined to describe this view. Physically, energy is transmitted down from large scales to smaller scales until dissipation through viscosity and heat occurs. There is a well known parody due to Richardson of Jonathan Swift's sonnet concerning fleas: "big whorls have little whorls that feed on their velocity, and little whorls have lesser whorls and so on to viscosity" which admirably and succinctly describes the process. Turbulence can therefore be considered as consisting of a continuous range of eddy sizes. These eddies are fed by many sources. Tides are a particularly good source in estuaries and coastal waters, but there are others such as wind and solar energy. Energy will be input at various length scales. What is evident is that most of the important effects of turbulence occur well after the source that input the original energy has ceased to have any direct influence. Turbulence is a subject worthy of a substantial textbook by itself. The ones by Hinze (1975) which is a classic, and the more recent Lesieur (1993) are particularly recommended. Let us proceed here by discussing turbulence in terms of variance, if only to allow easy comparison to the previous Fickian and shear models. If x is the separation of two particles of contaminant at time t, and u is the relative speed then from the definition of variance the following relationship holds between the variance of x at time t and its value at time $t = 0$:

$$\overline{x^2} = \overline{x_0^2} + 2\int_0^T \int_0^{t'} \overline{\delta u(t)\delta u(t+\tau)}d\tau dt'.$$

The integral is a measure of agreement called the *autocovariance* of a signal with itself. This idea will be met again in a different context entirely in Chapter 6. Some idea of its nature can be gained from knowing that it is zero for white noise where there is no relationship between a given value of the signal and any other no matter how adjacent the second might be. It is constant for a deterministic function. This way of looking at diffusion is due to L F Richardson and the function that is a locally based variance is sometimes called a Richardson distance neighbour function.

If the diffusion time T is very small the integrals can be dispensed with and, very nearly

$$\overline{x^2} = \overline{x_0^2} + \overline{(\delta u)^2}T^2.$$

This corresponds to the particle's trajectories being straight lines. For very large times, the particles would be so far apart that they would behave individually. The following bit of reasonably simple mathematics sorts out what will be the result in this case. Let

$$\delta u = u_2' - u_1'$$

so that

$$\overline{\delta u(t)\delta u(t+\tau)} = \overline{(u_2'(t) - u_1'(t))(u_2'(t+\tau) - u_1'(t+\tau))}$$

from which the right hand side becomes

$$\overline{u_2'(t)u_2'(t+\tau) + u_1'(t)u_1'(t+\tau) - u_1'(t)u_2'(t+\tau) - u_2'(t)u_1'(t+\tau)}.$$

In this expression, the last two terms are zero because the particles are so far apart that their speeds can be assumed to be uncorrelated. The first two terms are the same as there can be no distinction between these autocovariances. For, by the very nature of autocovariance, from its definition there can be no distinction as over a long time they will statistically speaking have the same history. Hence we have derived:

$$\overline{\delta u(t)\delta u(t+\tau)} = 2\overline{u'(t)u'(t+\tau)}$$

whence

$$\overline{x^2} = \overline{x_0^2} + 4\int_0^T \int_0^{t'} \overline{u'(t)u'(t+\tau)}d\tau dt'.$$

In Chapter 2 dimensional analysis was introduced. In particular the Buckingham Pi theorem gave a method by which to infer functional relationships between variables. This can be usefully employed here. Let us assume that the time rate of change of x^2 suitably time averaged depends solely on the initial separation x_0, the diffusion time T, the rate of transfer of turbulent energy ϵ. Applying the Buckingham Pi theorem, we see that the combination

$x_0 \epsilon^{-1/2} T^{-3/2}$ is dimensionless as ϵ has dimension $L^2 T^{-3}$. (In this last expression L and T are general length and time scales.) The view taken here is that of Kolmogorov in that the time rate of change of x^2 suitably time averaged does not depend on u the separation speed in the interesting range. This intermediate range which is between the initial linear range and the range when every particle has a very large separation is called the inertial sub-range. The deduction from dimensional analysis is that

$$\frac{d}{dt}\overline{(x^2)} = \epsilon T^2 F(x_0 \epsilon^{-1/2} T^{-3/2}).$$

Where the function F is general. For those relatively new to this game, time derivatives of $\overline{(x^2)}$ may seem a contradiction for doesn't the overbar imply a time average? The answer is that the time average is taken over a time scale appropriate to turbulent eddies whereas the outer time derivative refers to longer times. It is like a moving average that slowly varies with time. Some special cases deserve individual consideration and are illuminating. If T is small, we return to a linear separation rate and so

$$\frac{d}{dt}\overline{(x^2)} \propto T(\epsilon x_0)^{2/3}.$$

For intermediate values of T we can assert that there will be no dependence on the initial separation x_0 and dimensional analysis thus forces the function F to be a constant from which

$$\frac{d}{dt}\overline{(x^2)} = \epsilon T^2.$$

Direct integration shows that for small T

$$\overline{x^2} \propto T^2,$$

and for intermediate T we have

$$\overline{x^2} \propto T^3,$$

and for large values we return to

$$\overline{x^2} \propto T.$$

To summarise therefore we have a power law for the variance of a diffusing patch of contaminant:

$$\overline{\sigma^2} \propto T^p.$$

If T is large, $p = 1$ and we have the standard Fickian diffusion. For intermediate values of T, $p = 3$ which in the light of the previous section on shear diffusion is consistent with the interaction between "shear" and the concentrate (pollutant) producing enhanced faster diffusion rates. For very small values of T, $p = 2$ and particles are simply moving away from each other in straight lines. Interestingly, since

$$\kappa = \frac{1}{2}\frac{d}{dt}(\sigma^2)$$

it can be shown that from the case $p = 3$ (intermediate T; most rapid diffusion) we can derive $\kappa = \sigma^{4/3}$ known as "Richardson's four-thirds law". In fact Richardson proposed independently (Lewis Fry Richardson (1881 - 1953) was a *very* independent original thinker) that it was not really possible to distinguish between the flow and turbulent fluctuations. Richardson also deduced that the time over which quantities are averaged was crucial. The conclusion drawn was that the number of particles in a unit length, n obeyed an equation

$$\frac{\partial n}{\partial t} = \frac{\partial}{\partial l}\left[K(l)\frac{\partial n}{\partial l}\right]$$

where l is the standard deviation of the particles from their mean position. κ is a Fickian diffusion, but dependent on l. It is only after considering a large range of length scales that Richardson deduced the following functional form for K:

$$K = 0.2l^{4/3}.$$

That this is consistent with Kolmogorov isotropic turbulence theory for intermediate times reinforces both theories. Some further dimensional analysis is useful here. The turbulent kinetic energy spectrum $E(k,t)$ tells us how much turbulent kinetic energy is present at any given wavenumber and has dimension L^3T^{-2}. At these equilibrium ranges, the spectrum will only depend on the wave number k and the rate of dissipation of energy, ϵ. Accepting this, we can use dimensional analysis given k has dimension L^{-1} and ϵ has dimensions L^2T^{-3}. As only two variables are involved the elementary form of dimensional analysis can be used, that is the relating of powers without grouping variables into dimensionless clusters. Let

$$E \propto k^\alpha \epsilon^\beta$$

in dimensional terms only of course. As only L and T are involved, this is

$$L^3T^{-2} = L^{-\alpha}(L^2T^{-3})^\beta$$

$$\begin{aligned}\text{giving} \quad 3 &= -\alpha + 2\beta \\ \text{and} \quad -2 &= -3\beta.\end{aligned}$$

The solution of this is

$$\beta = \frac{2}{3} \quad \text{and} \quad \alpha = -\frac{5}{3}.$$

The result is the turbulence law

$$E = C_k \epsilon^{2/3} k^{-5/3},$$

which is known universally as Kolmogorov's $-5/3$rds law. This law, the Richardson four thirds law and the t^3 dependence of variance are all consistent and give an accepted picture of turbulence at intermediate scales (the inertial sub-range).

There are many observations that back up these laws, see Okubo (1974) for one example and the text by Lewis (1997).

This model due to Kolmogorov is called similarity theory. This is not to be confused with the use of *similarity variables* which is a technique used to obtain the exact solution to certain kinds of partial differential equation, amongst them the diffusion equation indeed. Similarity variables are used extensively in aerodynamics, but not here. Here the word similarity arises from assuming that diffusion is homogeneous in all directions. We have indicated that similarity theory gives variance proportional to t^3. Okubo reasons that the diffusion in the oceans is close to homogeneous, but constraints such as the ocean surface and perhaps the thermocline restrict the spreading. In fact some of the data he presents do locally fit a t^3 power law. Okubo's $t^{2.3}$ power law is still accepted today as the best simple law of oceanic diffusion. Okubo's actual relation is

$$\sigma^2 = 1.08 \times 10^{-6} t^{2.34}.$$

Since the diffusion coefficient κ is related to σ^2 through the relationship

$$\kappa = \frac{\sigma^2}{4t} = 2.05 \times 10^{-7} t^{1.34}$$

we have a power law for the diffusion coefficient too. Interestingly, drawing a regression line through the original data but this time plotting κ against l gives the relation

$$\kappa = 2.05 \times 10^{-4} l^{1.15}.$$

This can be compared to the theoretical law due to Richardson, the $l^{4/3}$ law.

Let us turn next to a different method of modelling diffusion that theoretically is quite old, but practically has only been of much use in the last fifteen or so years since the advent of high speed computers that are freely (in the sense of being more widespread!) available and which produce good graphical output. These models also make use of some of the ideas introduced in this section, but have a very distinct *modus operandi*.

5.2.4   Particle Tracking

In the 1970s, with the political crises in the Middle East and a general focus on saving energy, the world's attention on oil transportation by tankers intensified. Coincident with this, was an increase in public awareness of matters environmental. The final piece of the jigsaw was the unfortunate incidence of several oil spillages. Around the UK, the worst of these were the wreck of the *Torrey Canyon* in 1967 and, 11 years later, the break up of the *Amoco Cadiz*. There were many lesser spillages between these dates. The early 1970s also witnessed the beginnings of the computer revolution, making available a vast increase in computing power to modellers. The pressure to produce models of oil slick behaviour was therefore great.

A good method to model the behaviour of oil slicks, or any waterborne pollutant for that matter, is to hold the pollutant as a number of marked particles.

The position of each of these particles is individually held in the computer, and the power of the computer is such that all these positions are held at each time step, being updated by the model. In this way the evolution of a patch of pollutant can be tracked. Commercially produced software began in the mid-1970s; for example, one called SLIKTRAK was produced by one of the major oil companies. Since then, more and more sophisticated oil spillage tracking models have been produced. There is a tendency to call these models "random walk models" and to a large extent the name is a good one. In the simplest of models the marked particles are assumed to behave exactly like tiny bits of fluid. Hence the equations obeyed by the sea (the conservation of mass and momentum) are imposed as normal on all the fluid including the marked particles. In addition however these particles are also subject to local turbulence and so are pushed and pulled around by the eddies. If it is assumed that this turbulence has the characteristics implied by Fickian diffusion, then this can be simulated by assigning a random number by which to move a given particle off the line dictated by the local fluid velocity. If the number of particles is large enough, the result will be that the marked particles are distributed about the mean according to a Gaussian distribution. The variance of this distribution is dictated by the user and is related to the diffusion coefficient. Specifically the mean square deviation of the particles, $\overline{x^2}$ is related to the value of the diffusion coefficient κ_x and the time elapsed since the process began T which must be assumed large through the formula

$$\overline{x^2} = 2\kappa_x T.$$

Of course, the advection that took a great deal of attention to capture theoretically in the last two sections is there automatically simply as a result of the equations obeyed by the fluid (sometimes generally called the *primitive equations*, although this name should be preserved for constant density formulations only). So the net velocity of any given particle will be the vector sum of the local fluid velocity u and the displacement assigned by the random number divided by the time step of the numerical procedure used to implement the random walk. If this time step is Δt then using the above equation the displacement of a fluid particle is $\alpha\sqrt{2\kappa_x\Delta t}$ where α is the random number. Various modifications of this model can be used to simulate for example three dimensional diffusion. In this particular case, a three dimensional grid is defined and the movement of a marked particle within this box dictated by three random numbers corresponding to the three co-ordinate directions. A continuous or intermittent discharge can similarly be simulated by allowing new marked particles to enter at specified times and places. The method looks to be very powerful indeed. The only drawback being that a random number method can only hope to produce a spread whereby the variance of the patch grows linearly with time. Having said this, of course enhanced diffusion usually results from shear present in the numerical solutions of the primitive equations to which the particles are being subjected. In the final analysis, all that is missing is the interaction between eddies and flow that Richardson in particular insisted characterised turbulent diffusion in the sub-inertial range. We will later look at the implementation of one of the first random walk models.

A relatively straightforward particle tracking model, one of the first practical ones was produced by Hunter (1980), and it is this model that is outlined as a case study later. It is worth pointing out that the seductive idea of being able to move particles around in a manner that seems so closely to mirror reality has been explored by several other researchers. Dyke and Robertson (1985) proposed a theoretical turbulence which was composed of randomly distributed and randomly sized eddies. They showed that the variance of a patch of oil in such a flow grows as time to the power of between 1.8 and 3. This is consistent with Okubo's result, but problems of initialising and calibrating such a model make it rather impractical for most purposes, Jenkins (personal communication) uses a model based on waves. The waves are random in direction with correlation times differing by factors of 2. Jenkins reproduced the Kolmogorov t^3 law for the increase in variance. It is true therefore that there remains much scope in models of this kind. Perhaps one day, such a model will give a true picture of turbulent diffusion. A cautionary note, however: complicated models that involve many parameters can, chameleon like, be made to simulate any set of data without improvement of understanding. Some biological models but not those of Chapter 7 demonstrate this all too well.

5.3   Box Models

One of the earliest ways of modelling diffusion in estuaries and bays is to use box models. The basic method is to divide the estuary or bay into boxes, usually rectangular or square regions. Normally, attention was restricted to two dimensional models, mostly because two dimensional and one dimensional models were all there were back then. These two dimensional box models were either depth integrated area models over the horizontal or a vertical section along the centre line of an estuary. However in principle there is no reason why the box model concept could not be used for three dimensional modelling. Once the region is segmented into boxes then in any given box all variables are taken as constant. This is equivalent to averaging quantities over each box. The box model itself is built up using continuity of variables. What goes into a box through adjacent boxes must equal what emerges from other adjacent boxes. This could be done for salinity, current, temperature and other chemical and biological variables. The network of boxes are a means of "modelling" given data and there is really no modelling in the numerical sense, nor some would say in any other sense either! Certainly it is only extremely crude. However, there is a parameter which is still quite a favourite and sought after amongst coastal engineers working with estuarial pollution and that is the "flushing time". This is the time taken for any given foreign matter present in the estuary or other semi enclosed environment to exit into the open sea. Flushing time was commonly estimated using box models by estimating the time fluid particles take to cross each box and then adding these times together. The worst case is the longest possible time and this is an upper bound for the flushing time. However box models have largely been superseded by numerical models. In fact numerical models based on finite difference grids are very fine scale computationally based box models, except that the variables at the grid points are based on the equations not on measurements (except perhaps at the open boundaries).

5.4 Case Studies in Diffusion

5.4.1 A Particle Tracking Model

As we have already stated, good method to model the behaviour of oil slicks, or any waterborne pollutant for that matter, is to hold the pollutant as a number of marked particles. The position of each of the particles is individually held in the computer, and the power of the computer is such that all these positions are held at each time step, being updated by the model. A relatively straightforward model, produced by Hunter (1980), is now outlined to give the general flavour of these particle tracking models.

Hunter's aim was to produce a computer model that could simulate the behaviour of an oil slick lying on the surface of the sea. Hunter also wished to develop computer software that could be used by an unskilled operator. To this end, the software would have an easy to use menu-driven front end so that anyone who had an interest in oil slick movement, but who was not perhaps familiar with the details of modelling, could input data and interpret output. This is now quite a common feature, but it was not back in 1980.

When creating a model like this, one needs to decide what effects to include and what to exclude. This model predicted the movement of surface oil slicks, so the following factors were assumed to be important:

1. The tidal or non-tidal motion in the underlying seawater;

2. A wind-driven motion localised at the air-sea interface and caused by such mechanisms as surface water velocity, surface gravity waves and the direct action of the wind stress on the oil slick;

3. Spreading and mixing processes (eg. gravitational and surface tension effects, surface wave activity and horizontal turbulence.)

The underlying water movement, encapsulated in (1) and (2) above, causes the sea currents. These currents cause the patch of oil to move around as a whole, as well as to distort. However, the principal effect is to move the centre of mass of the oil slick. It is the spreading and mixing processes (3) that cause the slick to diffuse, i.e. to increase in size.

The tidal and other water movements are obtained, not by the model, but by interpreting observations. Other, later oil slick models have been entirely model based, using as their database the currents predicted from a primitive model of the type outlined in Chapter 3. Hunter's model however simply uses a file of current data from Admiralty charts and other sources. In order to supply the wind-driven data (2) an empirical formula relating wind speed and direction is

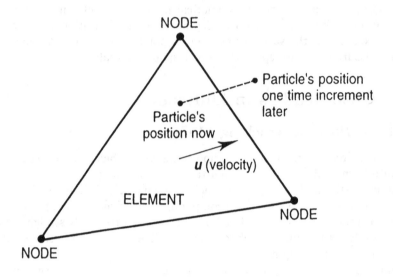

Figure 5.6: A typical element

used. If a more sophisticated wind-driven current were to be introduced, some account would have to be taken of the rotational effects of Coriolis acceleration. Indeed, one of the suggested improvements by Hunter himself is to include some kind of Ekman effect. In this simple model however, all such effects were omitted in favour of a direct drift that solely moves the centre of mass of an element in the direction of the wind but at some fraction of the speed. The name *element* here refers to one of the triangles into which the domain of the slick is divided to facilitate numerical computation (see Chapter 3). It is a novel treatment of (3) that deserves most attention here. In order to simulate spreading by diffusion, Hunter incorporates a so-called Monte Carlo technique. This is a technique whereby a particular particle is moved randomly in a direction with a particular speed. Thus if one imagines a speeded-up film of a bird's-eye view of an oil slick composed of such particles, a kind of Brownian motion would be observed with individual particles dashing madly about, but with the whole patch growing steadily in size. This has a comforting feel of realism about it, but unfortunately only Fickian diffusion can be modelled by a random walk. It can be shown that under a random walk, the variance of the distribution of oil in an oil slick has to vary directly with time t and not, for example with $t^{2.3}$, as Okubo's data indicate might be the case. In his model, Hunter divided the domain into triangular elements (in the manner of a conventional finite element method), as shown in Figure 5.6. The particles of oil are tracked across the triangular mesh using an implicit finite difference scheme. The position of each particle is thus continually updated. The underlying velocity due to tides mostly, is taken as a constant throughout the element, as is the wind drift effect. Omitting the computational details, an analysis of the diffusive spreading process is given. The example of a radially symmetric patch after n applications of the finite difference scheme gives a variance of:

$$\frac{(\text{speed} \times \text{time increment})^2 \times n}{2}.$$

Putting $t = n \times$ time increment and writing s as the distance travelled by a particle in one step gives variance $= ust/2$, as required by theory (and a Fickian model). Hunter's model also catered for the addition and removal of particles. It could not however, cope with coastlines or with oil that sank or differentiated into several different types of oil with different characteristics. Hunter's model was distinctive in that it was the first to incorporate the user explicitly, and that was its principal value. It is recognised that the model itself was too simple to be widely applicable. Here is a summary of model inputs and outputs:

Inputs

1. Wind drift factor (proportion and direction).

2. Diffusivity in SI units.

3. Decay constant.

4. Number of particles to be released initially (typical values would be 10 to 100).

5. Initial patch size, for instantaneous release.

6. Release position in latitude and longitude.

7. Time increment for computation (typically 900 s).

8. Time of initial release in GMT.

9. Whether the particle checking is required within critical elements.

10. Elapsed time for next model output in hours, minutes and seconds.

11. Wind speed and direction for period defined in (10).

12. Release rate of particles from point source over period defined in (10).

outputs

1. Number of particles.

2. Coordinates of the centre of gravity of the patch.

3. The direction of the principal axis of the patch.

4. The standard deviations parallel to and normal to the principal axis of the patch.

Initially, the patch is an ellipse of particles which is inputted under input (5). an additional map output with representative coastlines was also possible. A flow chart indicating the process is given in Figure 5.7.

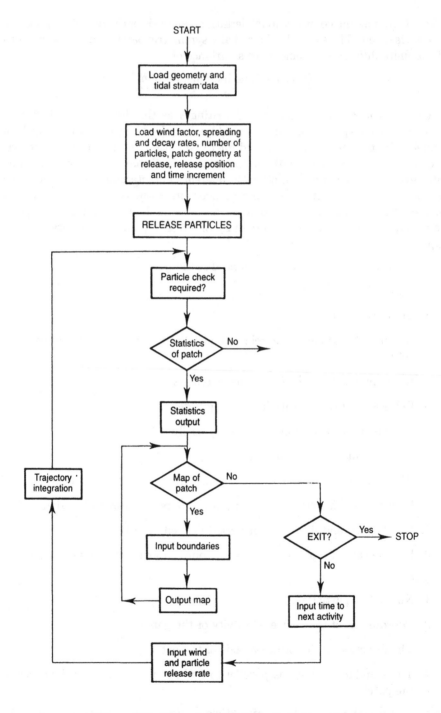

Figure 5.7: Flow chart for the oil slick model

5.4.2 Modelling diffusion in the North Sea

The motivation for the study of diffusion in the North Sea is tracking and monitoring of pollutants. In the late-1980s, a collaborative European study (the North Sea Programme) brought together researchers from countries that border the North Sea with the object of furthering our understanding of the underlying processes governing the behaviour of the sea itself, the sediments and the life within the sea. It was widely recognised that within this ambitious programme, one of the most important processes to get to grips with was diffusion.

The overall circulation pattern in the North Sea is reasonably well known and is free of controversy. There is an anticlockwise flow; southwards down the east coast of the UK, eastwards along the northern coasts of Belgium, The Netherlands and Germany, and then northwards along the Danish coast, and finally along the west Norwegian coast to exit the North Sea (Figure 5.8). There is some 'leakage' across the North Sea in the form of the Dooley current, which is an intermittent feature flowing from the Scottish-Northumbrian coast, across the North Sea, towards the Skagerrak. The model itself is based on a finite difference discretisation of the type described in Chapter 3 (some details of the finite difference schemes themselves are given in that chapter). On top of the numerical model of water movements need to be added the inputs from the Firth of Forth, the rivers Tyne, Humber and Thames in the UK, and from the rivers Rhine/Meuse, Elbe, Schelde, Seine and Ems from continental Europe. These rivers often contain foreign material in the form of discharges (industrial waste) and runoff from farmlands containing fertiliser, which are carried into the North Sea as pollution. The North Sea, being a semi-enclosed basin, is particularly vulnerable to environmental stress (or environmental impact as it is known nowadays).

These discharges are assumed to be in the guise of particles. The model is based on solving the momentum equations but in a form that uses particle tracking (Lagrangian) techniques, in order that the pollution can be tracked explicitly. The horizontal resolution, by which we mean the spacing between the grid points, is about 20 km, and the time step is three hours. Two models are in fact used: a two-dimensional model with the stated resolution and time step, in which all quantities are integrated through the vertical (depth averaged), and a more sophisticated model which has 10 levels in the vertical and a much smaller time step of 12 minutes. The results from the two models are not significantly different. In both models the time-dependent tidal current, not the mean, is used as the basis for computing the diffusion (spreading) of the particles, and in order to facilitate the handling of what would otherwise be too much data, the spreading simulations are vertically integrated. Spreading itself takes place within the model in two ways. The currents themselves exhibit turbulent fluctuations which by virtue of small eddies can spread passive contaminants (as represented by the 100 particles) that lie within the sea. In addition, there are mixing coefficients in the model that purport to represent directly the turbulent diffusion process. Other diffusion processes that are not directly hydrodynamic in origin, such as those due to biological or chemical agents are not present in

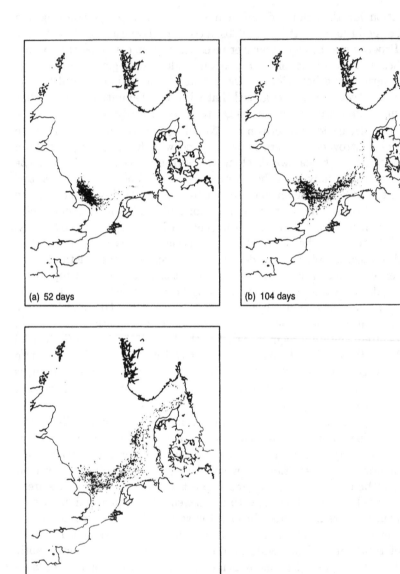

Figure 5.8: The diffusion of particles in the southern North Sea by diffusion and mixing

this model. There is an important distinction between the diffusion processes modelled here: *passive*, in which particles are carried around by ambient flow (or perhaps at some fraction of the ambient flow); and biological and chemical processes which, although they may also be largely diffusive, are also *active* in that internal biology and chemistry can take place. Biological organisms can also propel themselves of course! This kind of active modelling is the subject of Chapter 7. Let us finish with a more technical description of a case study; this one is an attempt to model the Gulf oil spill that was the result of the Gulf War of 1991. This is the largest known oil spill, may its like never be seen again.

5.4.3 Modelling the motion of spilt oil

The spill modelled outlined here follows Proctor, Flather and Elliott (1994). The first requirement of the model was a reliable tide and surge model of the type outlined in Chapter 6. This is the primitive equation model that is used to predict the underlying current. The model for tides and storm surges used the depth averaged equations, good enough in these circumstances. The difficulties of three dimensionality, stratification and turbulence closure are avoided. Here are the equations:

$$\frac{\partial \eta}{\partial t} + \nabla.(H\mathbf{u}) = 0$$

$$\frac{\partial \mathbf{u}}{\partial t} + \mathbf{u}.\nabla\mathbf{u} + f\mathbf{k} \times \mathbf{u} = -g\nabla\eta - \frac{1}{\rho}\nabla p_a + \frac{1}{\rho H}(\tau_s - \tau_b) + \nu_H\nabla^2\mathbf{u}.$$

The usual notation has been adopted, with the addition that H is the total water depth, p_a is atmospheric pressure an important feature to include for storm surge modelling, ν_H is the horizontal diffusion of momentum (eddy viscosity), and τ is the stress at the surface (s) and the bed (b). At the surface, a quadratic law connects the stress to the wind vector $\mathbf{W}$:

$$\tau_s = C_D\rho_a\mathbf{W}|\mathbf{W}|.$$

Here ρ_a is the density of air and the drag coefficient C_D is usually weakly related to the wind speed W. At the sea bed a similar quadratic law is assumed to hold, viz.:

$$\tau_b = \kappa\rho\mathbf{u}|\mathbf{u}|.$$

This time the drag coefficient $\kappa = 0.0015$ a constant value. Harking back to Chapter 3, an explicit finite difference scheme on a spherical but regular longitude - latitude grid was used. Obviously the authors used the tried and tested North Sea basic storm surge model refined by Roger Flather, the second of the authors of this Gulf oil spill paper. The open boundary condition used was

$$u_n = \hat{u}_n + \sqrt{\frac{g}{h}}(\eta - \hat{\eta})$$

where $\hat{\eta}$ and $\hat{u}_n$ are specified functions of space and time including contributions from both tide and surge. This enabled waves to propagate out of the sea area in a natural manner. The solid boundary conditions were the usual zero normal flow and the grid itself was 5′ by 5′ or approximately 9km. square. An additional complication was that the model required ten tidal constituents to get the tides sufficiently accurate. Meteorological data from a global model was used for wind and air pressure input, and the storm surge plus tide model validated at two ports. This then gave a sound basis upon which to add the spill model itself. The spill model used in this case is based on random walk modelling and not the conventional advection-diffusion equation. The particles of oil are pushed and pulled around by current shear in the form of advection, turbulence which is manifest through diffusion (particle separation) and buoyancy which depends on size of droplet and oil density. The model also includes evaporation and decay through the water column (or should that be "water" column!). The actual random walk formula used enables both horizontal and vertical diffusion modelling to take place. If the diffusion coefficients are labelled D_H (horizontal) and D_V (vertical then:

$$D_H = R(12E_H\Delta t)^{1/2} \quad \text{in the direction} \quad \theta = 2\pi R$$

and

$$D_V = (2R - 1)(6E_V\Delta t)^{1/2}$$

where R is a random number chosen between 0 and 1, E_H and E_V are horizontal and vertical diffusion coefficients and of course Δt is the time step. For small droplets (less than a critical diameter d_c obtained by matching droplet motion and considering Reynolds number) it is assumed that the rise velocity w can be described by

$$w = \frac{gd^2(1 - \rho_0/\rho)}{18\nu}$$

whereas for large droplets it is assumed that

$$w = \left[\frac{8}{3}gd(1 - \rho_0/\rho)\right]^{1/2}.$$

In these formula besides the droplet size d we have the viscosity of sea water ν and ρ_0 is the density of oil. The probability p of removal of a droplet at each time step is given by

$$p = 1 - e^{-(\lambda\Delta t)}.$$

In practice, for each droplet a random number R is generated at each time step and if $R \leq p$ the droplet is removed. The decay constant λ is related to the time scales of evaporation and degradation. Obviously only oil droplets close to the surface experience evaporative decay but oil throughout the water column experiences degradation. Using standard properties of hydrocarbons, typically in less than 10 days most of the oil that is going to evaporate (25%) has done so and that about 5% has degraded. This allows a decay or "e-folding" time which is the reciprocal of λ to be determined. The droplets are of course also moved horizontally and if they are either beached (that is hit a solid boundary)

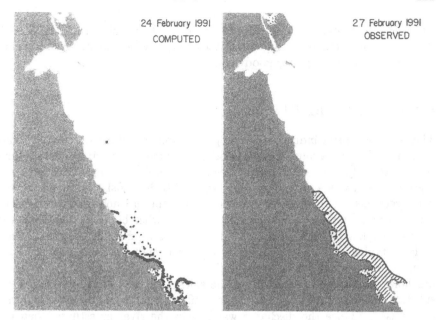

Figure 5.9: Oil slick prediction and observed position

or move out from the open boundary then they take no further part in the simulation. The paper by Proctor, Flather and Elliott (1994) then goes on to discuss the specifics of a particularly large oil spill in the Gulf in January 1991. The data used was $\rho_0 = 870 kg\, m^{-3}$ $E_H = 10 m^2 s^{-1}$ $E_V = 0.005 m^2 s^{-1}$ and Δt is one hour. The droplets of oil are between 60 and 120 μm. Simulated and observed results are reproduced in Figure 5.9. The result of the simulation was broad agreement between model and observation which was gratifying. It is unusual for such a comparison to be possible. Deliberately causing a slick then monitoring it to validate a model is not an option. All of us would rather the Gulf oil slick did not happen, but taking a small grain of comfort from such an ecological disaster it is good to test a sophisticated spill model and have it vindicated. More details including a further account of shortcomings are given in the original paper. Interestingly, it is thought that even given the disastrous short term pollution caused by the vast spill, the long term pollution would be significantly decreased by the decrease in tanker traffic during the Gulf War itself!

Finally, the paper by Elliott (1991) outlines the model EUROSPILL which includes the kind of droplet behaviour indicated above together with an accurate numerical tide and surge model. Wind shear is included through a parameter which represents the thickness of the wind shear layer. An additional mechanism not before mentioned is the Stokes drift due to the non linear interaction of surface waves. These give rise to a drift (see Chapter 6) that can be significant. Diffusion is, of course present as is buoyancy. The really novel aspect of EUROSPILL is the inclusion of a user friendly menu and associated graphics which enable a novice user to drive the model. The usual warning applies; the glamorous output is only as good as the modelling behind it. The modelling behind this one looks pretty good though.

5.4.4 Modelling Plumes

The civil engineering industry requires plume models that are valid for rivers and estuaries. This is because many factories are cited beside rivers and estuaries so that the adjacent water can be used perhaps for cooling and also so that there can be a waste discharge. Since the 1970s the legislation governing such discharges has become much more severe. Europe has now also got in on the act. There are long lists of chemicals that are either banned or are such that only a very low level of concentration is permitted. Turning to what actually is done, typically there is a discharge pipe that ends part way into a river or estuary, and through which the material flows. Usually, this is not a pipe but has holes strategically placed along the sides to maximise the dilution of any substance that might be present in the outflow material. This device is termed a "diffuser". Once the discharge water is in the river or estuary, then it is important to model its behaviour as accurately as possible in order to decide whether or not legislation is breached. The usual method is to consider the plume as a buoyant jet which diffuses in accordance with a Gaussian law. The diffusing jet is then subject to tide, wind, river flow all of which under normal circumstances help the dilution. The case of most interest is always the worst case scenario where all dilution mechanisms are least. Simple to model of course as zero flow due to river, wind or tide does the trick. The other environmental question to ask is whether any unwanted material is likely to find its way on to a nearby sensitive part of the river or estuary, an SSSI (Site of Special Scientific Interest) perhaps, a holiday beach or yachting marina. In all these questions modelling plays a pivotal role. However, let us turn to a larger model for a detailed case study; the Rhine river plume as it exits into the North Sea.

The Rhine is the largest river discharge into the North Sea and therefore there is good reason to model it. The model presented here is a summary of the paper by Ruddick *et al* (1994). The initials ROFI (Region Of Freshwater Influence) join the plethora of acronyms, this set being employed to indicate the waters around the mouth of the estuary that are stratified and otherwise influenced by the river discharge. The model used is not particularly simplified but employs three dimensional hydrodynamics with stratification incorporated via buoyancy and a conservation of salinity. The momentum equations themselves use only the constant reference density in the pressure term; this is called the Boussinesq approximation. The momentum equations employ an eddy viscosity, and the conservation of salinity a diffusion coefficient. Some of the symbols used are different to those introduced here, so to give the equations of the paper verbatim would confuse, suffice it to say that the horizontal momentum conservation equations contain non-linear terms, Coriolis terms and eddy viscosity terms. Instead only certain differences will be highlighted. The buoyancy b is given by

$$b = \beta_S g(S_0 - S)$$

where g is the acceleration due to gravity, β_S is the coefficient of haline expansively and S_0 is a reference salinity. A reduced pressure is defined by

$$q = \frac{P - P_{atm}}{\rho_0} + gz$$

and the buoyancy is then

$$b = \frac{\partial q}{\partial z}.$$

The conservation of salinity is the standard advection-diffusion equation:

$$\frac{\partial S}{\partial t} + \frac{\partial}{\partial x}(uS) + \frac{\partial}{\partial y}(vS) + \frac{\partial}{\partial z}(wS) = \frac{\partial}{\partial z}\left(\lambda_S \frac{\partial S}{\partial z}\right),$$

this equation, and the momentum equations not quoted here, are written in what is called "flux" form. (Nothing too mysterious here; simply take the continuity equation:

$$\frac{\partial u}{\partial x} + \frac{\partial v}{\partial y} + \frac{\partial w}{\partial z} = 0$$

multiply it by S and add to the standard form of the advection diffusion equation. The combinations

$$u\frac{\partial S}{\partial x} + S\frac{\partial u}{\partial x} \qquad \text{etc.}$$

then appear which are combined to give the fluxes uS etc. differentiated.)

The Arakawa C grid is used and the variable depth domain transformed via the sigma co-ordinate (see Chapter 3) into lying between two constant values. It is not the purpose of this chapter to delve into another three dimensional hydrodynamic model, albeit one with novel features, this belongs in Chapter 3. Instead let us look at features that pertain to the river plume modelling. Boundary conditions were dealt with in the last chapter, and in particular what boundary conditions are appropriate to impose at the open boundaries, and there are three of them here. Figure 5.10 shows the domain, and the three open boundaries. The model has to allow for tide, wind stress and river as forcing mechanisms, therefore the tide must be allowed to enter and leave the model area. Here are the boundary conditions imposed. They will be explained once they have been stated.

$$U + c\eta = 2c\eta_0 \sin \omega_0 t, \qquad (5.1)$$

$$U - c\eta = 0 \qquad (5.2)$$

$$V - c\eta = c\eta_0 \sin\left(\frac{\omega_0 x}{\bar{c}} - \omega_0 t\right), \qquad (5.3)$$

$$V + c\eta = \frac{2}{A}(Q_0 - Q_1 \sin \omega_0 t). \qquad (5.4)$$

Taking the symbols in the order they appear, U is the alongshore transport (units $m^2 s^{-1}$), V the cross shore transport, $c = \sqrt{gh}$ is the wave speed (see

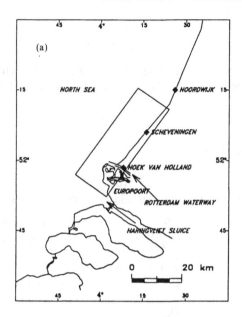

Figure 5.10: The Rhine plume model study area

Chapter 6) for long gravity waves, h is the mean water depth, η is the surface elevation, $\omega_0 = 1.405 \times 10^{-4}s^{-1}$ the frequency of the tide (M_2 - semi diurnal lunar tidal constituent). The value $\eta = 0.9$ is taken as a typical amplitude for the tidal forcing, and $\bar{c} = \sqrt{20g}$ a typical wave speed based on a depth of $20m$. A taken as $1km$ a representative width of the river mouth, and finally $Q_0 = 1200m^3s^{-1}$ and $Q_1 = 1800m^3s^{-1}$ are constant tidal components of vertically integrated discharge. These conditions are imposed at the two boundaries that cross the shore. At the outer offshore boundary a radiation condition is imposed (see Chapter 4) which here takes the form:

$$\left(\frac{\partial}{\partial t} + c\frac{\partial}{\partial y}\right)(V + c\eta) = -c\frac{\partial U}{\partial x} + \tau^{surf} - \tau^{bed} - fU - \hat{A}^h + \hat{\mathcal{L}}_d \qquad (5.5)$$

where τ^{surf} and τ^{bed} are the surface and bottom stress and the terms $\hat{A}^h$ and $\hat{\mathcal{L}}_d$ are the vertical integrals of horizontal advection of momentum and internal pressure gradient respectively. The idea is to compute $V - c\eta$ using equation 5.3 and $V + c\eta$ using equation 5.5. What equation 5.5 is designed to do is to allow the internal solution to develop naturally. Reflection at the open offshore boundary is further discouraged by imposing the condition that the normal gradient of the deviation from vertically averaged horizontal speed in the direction of the normal is zero. In symbols:

$$\frac{\partial}{\partial y}\left(v - \frac{V}{H}\right) = 0.$$

Salinity needs careful treatment too; it is carefully allowed to leave and enter at

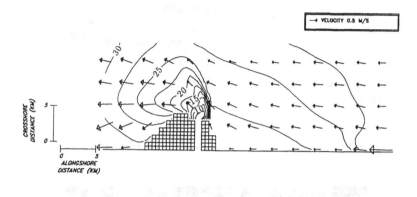

Figure 5.11: The Rhine plume: surface currents and salinity at low tide

at open boundaries by using upwind differences (see Chapter 3) and by supposing that incoming fluxes correspond to advecting the reference salinity S_0. At the river discharge, a two layer situation ($5m$ upper layer, $25m$ lower layer) is applied. At the river, the flux enters the domain. Given the uncertainty of the modelling assumptions that surround the open boundaries, validation and verification studies are essential. The authors were encouraged to apply the model to real situations thanks to an accurate Kelvin wave simulation (for more about Kelvin waves, see the next chapter). Figure 5.11 gives one output graph from the model. The next picture Figure 5.12 gives a schematic alongshore transect. The freshwater jet is indicated spreading into the offshore tidal stream. It remains identifiable as a buoyant jet and stays close to the surface of the sea. This is a complex area to model, and this model of Ruddick *et al* (1994) is certainly not the final word on the subject. Sharp fronts occur in reality and cannot be resolved using this model. Arguments about the use of diffusion coefficients for plume dynamics must be listened to. Unrealistically large diffusion would render the model useless for water quality purposes. Variations in river flow have also been ignored and these are patently a very important aspect of the plume. It is interesting to compare the kind of results given in Ruddick *et al* (1994) with the kind of model outlined earlier where the outfall water is marked and tracked using the diffusion equation or a more statistically based approach. In fact the very next paper in the journal after Ruddick *et al* (1994), de Kok (1994) describes such a model. de Kok solves the advection diffusion equation, but only in two dimensions. The novelty lies in a complex uneven dispersion mechanism that seems to give realistic patterns of discharge, see Figure 5.13. Neither type

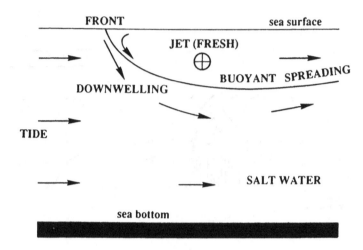

Figure 5.12: A Schematic alongshore transect through the plume

of model is able to reproduce fronts. In order to do this the numerical techniques outlined in Chapter 3 need to be employed.

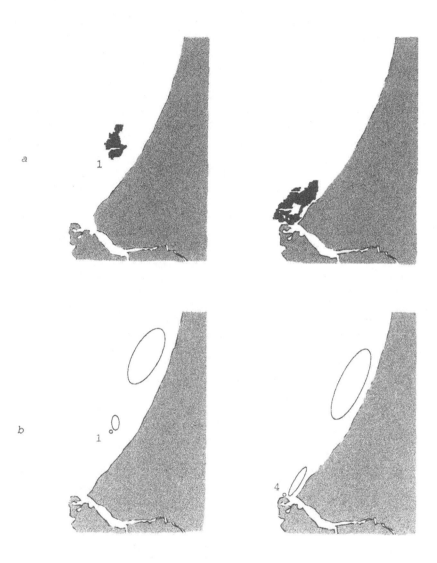

Figure 5.13: Dispersion patterns, from de Kok (1994)

Chapter 6

Modelling on the Continental Shelf

6.1 Introduction

Oceans have a general structure which is dictated by their geographical origin. There are mid-oceanic ridges, trenches and, most important for us here, continental shelves. These shelves border the margins of the oceans, the seas over them are only about 200 m deep, and the transition from this 200 m to the oceanic 3000 - 4000m is achieved through a relatively small region, the continental slope - which is aptly named, since the slope is commonly quite mild, only 4°. Simple trigonometry shows that the depth can sink from 200 m to 3000 m in a horizontal distance of 40 km with the slope. Since the width of the ocean is measured in thousands of kilometres, and most continental shelves are hundreds of kilometres wide, this slope region is of insignificant horizontal extent. It is of course, only the dimension that is insignificant. The continental slope contains important currents and is a significant source and sink of energy for many different types of flow. However, importantly for continental shelf modellers it serves to mark the border between deep ocean and continental shelf, and is the site of the open boundary condition. The dimensions of the continental shelf render models of continental shelf seas quite distinct from ocean models. In this chapter we shall be looking at modelling the waves and currents that occur on the world's continental shelves and also the waves that exist and may be trapped on the continental slope. Let us start with surface water waves.

Surface water waves are a very important part, one might say essential property of the sea. Waves are what we see first when we look at it, and they certainly have a great influence on man's interaction with it. However, there are many different kinds of waves that occur in oceanography. The surface water waves that finish their existence by crashing on to the beach are but one type. Tides are waves, as are meanders of the Gulf stream and the oscillations that accompany the phenomenon that takes place in the Southern Pacific called El Niño.

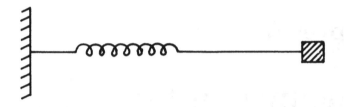

Figure 6.1: A mass connected to a spring

What we shall do here is give a short account of the general terms needed to describe all waves, then go on to consider models of waves that occur in continental shelf environments. If you know something about waves already, then you may well be able to skip the next paragraph.

6.2   Modelling Waves

The essential ingredient common to all waves is the to-and-fro motion, without there being any overall movement in any direction. Offshore engineers and oceanographers need to understand many things that oscillate; a few examples are ships, offshore structures subject to high winds and the sea itself. At first sight, all of these seem quite different, and very difficult to model. However, they all exhibit a to-and-fro motion, and it is this that we will attempt to describe. We will initially restrict our attention to motion in one dimension, or more strictly, to what engineers and applied scientists call a single degree of freedom system. Later, this restriction will be lifted as ocean scientists usually have to deal with two dimensional waves. Consider a mass attached to a spring on a smooth horizontal table as shown in Figure 6.1. If the spring is neither stretched nor compressed, the mass will not be subject to any force, and therefore it will be in equilibrium. If the mass is pulled (or pushed) in the line of the spring and then released, it will vibrate back and forth. The spring will always try to restore the mass to its equilibrium position, but it will overshoot and in the absence of any damping or friction, never come to rest other than instantaneously. Perfect springs have no damping, so the simple spring system shown in Figure 6.1, once

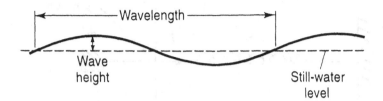

Figure 6.2: A surface water wave

set in motion, will vibrate for ever. Compare this with the water wave shown as Figure 6.2. This too exhibits the to-and-fro motion and it should come as no surprise that they are modelled in an identical way, at least initially.

In order to explain the terms used in describing waves, let us look at the mass spring system of Figure 6.1. The distance between the position of equilibrium where the mass has no forces acting on it and thus does not move to the furthest point reached by the mass in either direction of its motion is called the *amplitude* of the vibration or oscillation. The words *vibration* and *oscillation* are used synonymously. The time the mass takes between leaving the left-most (say) extremity and returning there is called the *period*, and 2π divided by this number is called the *frequency* of the oscillation. If a particle is describing a horizontal circle with uniform speed, the time taken to describe the circle once is the period (one cycle). If the perpendicular from the particle to the diameter is drawn, the foot of this perpendicular will describe *simple harmonic motion* (SHM), that is it will oscillate. The radius of the circle is the amplitude of the vibration. This provides a useful alternative view of vibration.

6.3 Simple Harmonic Motion

SHM is ideal in the sense that in order to describe the motion of springs the springs have to be well enough behaved to obey Hooke's Law whereby extension is proportional to the applied force. To derive an equation to describe it, we must use the second of Newton's laws of motion which is not really appropriate here

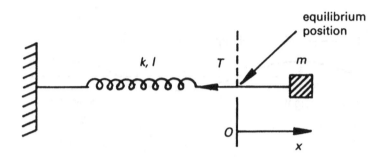

Figure 6.3: Defining some terms

Figure 6.3 is the same as Figure 6.1 but with the addition of an origin, axis, force and some labels. The equation obeyed by this mass spring system writing k for the stiffness of the linear spring is

$$m\frac{d^2x}{dt^2} = -kx. \tag{6.1}$$

Dividing by m and writing $\omega^2 = k/m$ yields:

$$\frac{d^2x}{dt^2} = -\omega^2 x \tag{6.2}$$

where ω is a number with dimensions of $(\text{time})^{-1}$ called the *natural frequency* of the mass spring system. This name arises from the fact that it is the frequency at which the mass oscillates when pulled to one side and released. We can see this because the solution to equation (6.2), a second-order differential equation with constant coefficients, is:

$$x = A\sin\omega t + B\cos\omega t \tag{6.3}$$

where A and B are constants obtained by using given conditions on x or dx/dt at specific times (usually at $t = 0$). An example should make the determination of A and B clear.

Example 6.1 *Calculate the displacement x in terms of ω and t if, at time $t = 0$, $x = 3$ and the velocity is zero.*

Solution The velocity of a particle that is displaced a distance x is its rate of change with respect to time, dx/dt so if x is given by equation (6.3), then differentiating with respect to t gives:

$$\frac{dx}{dt} = \omega A \cos \omega t - \omega B \sin \omega t. \tag{6.4}$$

At time $t = 0$, we obtain:

$$\left.\frac{dx}{dt}\right|_{t=0} = \omega A. \tag{6.5}$$

If this is to be zero, we must have $A = 0$.
Setting $A = 0$ results in the simplification of equation (6.3) to:

$$x = B \cos \omega t \tag{6.6}$$

and if $x = 3$ at $t = 0$:

$$B = 3. \tag{6.7}$$

We have thus determined the particular values of A and B that satisfy $x = 3$ and $dx/dt = 0$ at time $t = 0$. The solution is:

$$x = 3 \cos \omega t. \tag{6.8}$$

The introduction of simple sinusoidal waves through a mass spring system certainly has the advantage of being well defined. The analogy of the terms across to water waves works well provided the waves are considered sinusoidal. Unfortunately, water waves need not be sinusoidal, and they are extremely distorted as they approach a beach. Think about the wave that approaches the beach and spills over in the manner sought after by surfers everywhere. This is certainly not a sine wave. It can be shown that the manner in which a water wave begins to differ from being sinusoid is by a shallowing of the troughs and a steepening of the crests, but there are plenty of examples of waves in the ocean which are virtually sinusoidal (see the book "Waves in the Ocean" by P H Leblond and L A Mysak (1978)). Tides are one of these, and we consider tides a little later in this chapter.

In oceanography, waves can be quite complex and some more terms come in useful. For example, for waves that are not perhaps cleanly sinusoidal the *total height H* is a good concept. This is the vertical distance between the lowest trough and topmost crest of the waves. The *wave steepness* is the ratio H/L where L remains the *wavelength* which is the horizontal distance between successive crests (or troughs). The quantity $2\pi/L = k$ is called the *wave number*. The natural frequency (remember this is $2\pi/$period) of a wave in oceanography often tells us about the mechanism(s) that are being modelled. For tides, this would be a single tidal frequency or perhaps a sum of frequencies of several different tides. Adding up different waves in this way forms the basis of the harmonic theory of tides which still is a useful model. Harmonic analysis is a realisation of Fourier decomposition which might be familiar to some of you. It does not matter if it isn't. Another natural frequency for ocean dynamics at a horizontal scale of several kilometers is associated with the vertical component of the angular velocity of the earth and arises out of the Coriolis acceleration (see Chapter 2). This is the *Coriolis frequency* denoted by f where $f = 2\Omega \sin(\text{latitude})$. This frequency occurs all over the place in oceanography. If an iceberg is floating in a frictionless boundless sea, hardly realistic but ideal for us here, and is given a push and allowed to move freely it will travel in a circle of radius U/f where U is the initial speed imparted by the push. This circle is called an *inertial circle*, the velocity of the iceberg (u, v) is given by

$$u = U \cos ft, \qquad v = U \sin ft$$

and the circle itself by the cartesian equation

$$x^2 + y^2 = \left(\frac{U}{f}\right)^2$$

provided the direction of push is along the x axis and the co-ordinate system is appropriately chosen, (which means that the iceberg has co-ordinates $(0, -U/f)$ at time $t = 0$). This can be formally derived from the dynamical equations:

$$\frac{\partial u}{\partial t} - fv = 0$$

$$\frac{\partial v}{\partial t} + fu = 0$$

from which the SHM equations

$$\left(\frac{\partial^2}{\partial t^2} + f^2\right)(u, v) = 0$$

are easily obtained, and from which the above solution immediately follows. This SHM is called an *inertial oscillation*. The distance U/f is termed the Rossby radius of deformation and is a naturally occurring length scale for the width of coastal currents and upwelling provided the Coriolis parameter is assumed constant. In Chapter 2 the Rossby radius of deformation was seen as the natural width of the Gulf Stream under the balance between advective acceleration and Coriolis acceleration. Under the present assumptions, the advective acceleration terms happen to cancel to zero! That the same length scale emerges is interesting and can be deduced from the vorticity balance. What we have here is clearly a two dimensional application of oscillations. Truly two-dimensional sinusoidal waves are called *plane waves* and have the general form

$$a_0 \cos(kx + ly - \omega t + \phi) \qquad \text{or} \qquad ae^{i(kx+ly-\omega t)}.$$

Where the symbols k, l are wave numbers and the constant ϕ is a phase. In the second expression, only the real part has physical significance (a is complex, $a = a_0 e^{i\phi}$). It is convenient sometimes to adopt the powerful vector notation and write the plane wave in the form

$$ae^{i(\mathbf{r}.\mathbf{\kappa} - \omega t)}$$

where **r** is the two dimensional position vector ($\mathbf{r} = x\mathbf{i} + y\mathbf{j}$) and $\boldsymbol{\kappa}$ is the two dimensional wave number vector defined by $\boldsymbol{\kappa} = k\mathbf{i} + l\mathbf{j}$. The trouble some of you will have to go through to master this vector description will be well worth it if the need is to be able to deal with waves that are bending (due to refraction over topography perhaps) or being reflected obliquely. It opens up the use of wave ray theory. This notation also extends naturally to three dimensional "plane" waves which are not considered here. In the context of tides and waves on this kind of geophysical scale, these two dimensional plane waves are sometimes called Poincaré waves after the brilliant French mathematician Jules Henri Poincaré (1854 - 1912).

The speed at which a wave travels is labelled c and is called its wave speed or *celerity* and for purely progressive waves of the type $a_0 \sin(kx - \omega t)$ it is the ratio ω/k which is equal to wave period divided by wavelength. However for waves that are not purely progressive the wave speed needs to be found by other means. One property of waves yet to be alluded to but nevertheless important is *dispersion*. This is the tendency for a wave to decrease in amplitude as it travels. For all waves a relationship can be derived that connects the allowable wavelengths and wave frequencies. For us in the field of ocean science this relationship arises from considering the dynamics of water waves in a sea of constant depth h and takes the form

$$c^2 = \frac{g}{k} \tanh(kh)$$

where tanh is a mathematical function called the hyperbolic tangent found on scientific calculators and defined by

$$\tanh x = \frac{\sinh x}{\cosh x} = \frac{e^x - e^{-x}}{e^x + e^{-x}}.$$

For waves over very deep water, the depth h plays no part and very nearly

$$c^2 = \frac{g}{k}.$$

On the other hand, for shallow water the hyperbolic tangent is more or less a straight line so $\tanh x \approx x$ and this time the wave speed does not depend on the wavelength, instead

$$c^2 = gh.$$

These dispersion relationships can tell us a great deal about the behaviour of waves in various sea environments and are particularly useful for coastal and environmental engineers. In the real sea, one does not just have isolated sine waves, but a complicated surface of crests and troughs. Assuming that these can be approximated by a wave in one dimension then in turn they can be thought of as a sum of sinusoidal waves (a Fourier series in fact). Each component of this combination of sine waves will, according to the dispersion relation travel at different speeds. The simplest case is a wave train that consists solely of waves of two slightly differing wavelengths. Superimposing these will give "beats" of the type met when tuning stringed musical instruments. The beats themselves progress at a speed that is slower than the speed of either primary wave. If we label the speeds of the waves c_1, c_2, their wave numbers k_1, k_2 and their frequencies ω_1, ω_2 then it is easily shown that the speed of the "beat" is given by

$$c_g = \frac{\omega_1 - \omega_2}{k_1 - k_2}$$

which in turn is

$$c_g = \frac{c_1 \times c_2}{c_1 + c_2}.$$

If the speeds are very close, then very nearly

$$c_g = \frac{1}{2}c$$

where $c_1 \approx c_2 \approx c$. The speed c_g is called the *group velocity* and in general is the speed at which the energy of a group of waves travels. As said earlier, the group velocity of a wave train is always less than the celerity of any of the sinusoidal waves of which it is composed. In some cases, the group velocity can oppose the direction of the wave train! This is counter intuitive, but is true for Rossby waves (planetary waves) which have a group velocity that always has a westerly component. These waves are particularly interesting to study as they have unusual properties, however they are only found in the ocean as meanders in the Gulf Stream and currents of similar scale and therefore scarcely get a mention in a book on coastal sea modelling. The persistent westerly migration of energy lies behind the western intensification of ocean currents (Gulf Stream and Kuroshio).

Some of these basic notions of waves will be reinforced in the subsequent sections. Let us start with a look at continental shelf ocean dynamics.

6.4   Continental Slope and Shelf

There is an intriguing type of current where the change of depth at the continental slope and the fluid vorticity add, so that the movement of the current up and down the slope as it travels like a sidewinder along the slope itself ensures potential vorticity is conserved. The wave in this case is a current meander and there is little or no surface elevation associated with it. This is the continental shelf wave, which should really be called a continental slope wave. These waves were first predicted by Robinson (1964), and since the mid 1970s their importance to the computation of extreme currents in slope regions has meant continued interest and more sophisticated models. In this section the basic balances will be described in as simple terms as possible and complexities gradually introduced. The simplest equation set for motion on the continental slope is

$$\frac{\partial u}{\partial t} - fv = -\frac{1}{\rho}\frac{\partial p}{\partial x},$$

$$\frac{\partial v}{\partial t} + fu = -\frac{1}{\rho}\frac{\partial p}{\partial y},$$

$$\frac{\partial}{\partial x}(Hu) + \frac{\partial}{\partial y}(Hv) = 0.$$

If the pressure is eliminated from the first two by cross differentiation a potential vorticity balance is obtained of the type examined in Chapter 2. That is

$$\frac{\partial}{\partial t}\left(\frac{\partial u}{\partial y} - \frac{\partial v}{\partial x}\right) + f\left(\frac{\partial u}{\partial x} + \frac{\partial v}{\partial y}\right) = 0$$

and the continuity equation ensures that it is possible to define a streamfunction ψ such that

$$Hu = -\frac{\partial \psi}{\partial y}, \quad \text{and} \quad Hv = \frac{\partial \psi}{\partial x}.$$

Writing

$$\frac{\partial u}{\partial x} + \frac{\partial v}{\partial y} = -\frac{1}{H}\frac{\partial H}{\partial x}u$$

where it has been assumed that $H = H(x)$ the vorticity balance becomes

$$\frac{\partial}{\partial t}\left(\frac{\partial u}{\partial y} - \frac{\partial v}{\partial x}\right) = \frac{fu}{H}\frac{\partial H}{\partial x}.$$

Here it is assumed that the coast runs north-south and is west facing. Axes are y north and x east as usual and the equations are then in the same frame as those used to model upwelling in Chapter 3. With this simplification, the single equation in terms of the streamfunction ψ is

$$\frac{\partial}{\partial t}\left[\frac{\partial}{\partial y}\left(\frac{1}{H}\frac{\partial \psi}{\partial y}\right) + \frac{\partial}{\partial x}\left(\frac{1}{H}\frac{\partial \psi}{\partial x}\right)\right] - \frac{f}{H^2}\frac{\partial H}{\partial x}\frac{\partial \psi}{\partial y} = 0.$$

The way forward now is to look for wave like solutions that propagate in the y direction. If we let

$$\psi = H^{1/2}\phi(x)\exp(ily - i\omega t)$$

then it turns out that ϕ satisfies the ordinary differential equation

$$\frac{d^2\phi}{dx^2} + \left\{\frac{d}{dx}\left(\frac{1}{2H}\frac{dH}{dx}\right) - \left(\frac{1}{2H}\frac{dH}{dx}\right)^2 - l^2 - \frac{fl}{\omega}\frac{1}{H}\frac{dH}{dx}\right\}\phi = 0.$$

It is at this stage that actual profiles for $H(x)$ could be inserted into this equation, however the most instructive thing to do is to use the exponential depth profile

$$H = H_0\exp(-2\lambda x)$$

in which case the equation for ϕ simplifies considerably. The solution $\phi(x) = \phi_0 \sin kx$ is permitted and there is a dispersion relation

$$\omega = \frac{2fl\lambda}{k^2 + l^2 + \lambda^2}.$$

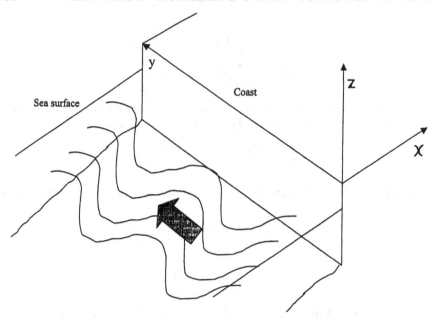

Figure 6.4: Continental shelf waves over idealised topography

Classical solutions in the paper by Buchwald and Adams (1968) are for shelf profiles that are exponential then flat:

$$\psi = \begin{cases} e^{\lambda(-x-B)}\sin kx e^{i(ly-\omega t)} & \text{for} \quad 0 \le -x \le B \\ e^{\lambda(x+B)}\sin kB e^{i(ly-\omega t)} & \text{for} \quad -x > B. \end{cases}$$

Figure 6.4 shows typical shelf waves. It may not be obvious, but these waves are quite large scale. The idealised geometry above contains the parameter λ the reciprocal of which can indicate the shelf width which is typically 30km. The frequency ω is always less than f, typically $\omega = 0.7f$ (for the case $k = \lambda$, Gill (1982)). In this case also, $f/\lambda = 3ms^{-1}$. So far nothing has been said about how these waves are forced. The above solution has been found simply by solving the basic unforced equations. It will come as no surprise to learn that it is surface wind stress that provides the forcing. The balance is expressed by the equations:

$$\begin{aligned} -fv &= -g\frac{\partial \eta}{\partial x}, \\ \frac{\partial v}{\partial t} + fu &= -g\frac{\partial \eta}{\partial y} + \frac{1}{\rho H}\tau^y(y,t), \\ \frac{\partial}{\partial x}(Hu) + \frac{\partial}{\partial y}(Hv) &= 0. \end{aligned}$$

In the above set of equations advantage has been taken of the previous unforced analysis to approximate longshore flow (cross shelf momentum of course) by geostrophic balance. The term $\frac{1}{\rho H}\tau^y(y,t)$ is due to the action of the wind stress, and only the component along the shore is effective in generating these waves and associated currents. This is reasonable given the geometry; the shelf is simply not wide enough for the cross shore component to grip the sea in any significant way. It is also supported by observational evidence. With these equations, progress via a streamfunction is possible as before, this time however a single forcing frequency is not appropriate as wind is seldom that helpful! The single equation for ψ is now

$$\frac{\partial^2}{\partial x \partial t}\left(\frac{1}{H}\frac{\partial \psi}{\partial x}\right) - \frac{f}{H^2}\frac{\partial H}{\partial x}\frac{\partial \psi}{\partial y} = -\frac{\tau^y}{\rho H^2}\frac{\partial H}{\partial x}.$$

The forcing at all sorts of frequencies means that a series of the form

$$\psi = \sum_{n=1}^{\infty} A_n(y,t)H^{1/2}\phi_n(x)$$

is sought which leads to what is termed an eigenvalue problem, namely:

$$\frac{d^2\phi_n}{dx^2} + \left\{\frac{d}{dx}\left(\frac{1}{2H}\frac{dH}{dx}\right) - \left(\frac{1}{2H}\frac{dH}{dx}\right)^2 - \frac{f}{c_n H}\frac{dH}{dx}\right\}\phi_n = 0.$$

In this equation c_n is the wave speed and the amplitude functions A_n satisfy the equation

$$\frac{1}{c_n}\frac{\partial A_n}{\partial t} + \frac{\partial A_n}{\partial y} = -\frac{b_n\tau^y(y,t)}{\rho f}$$

where the constant b_n emerges from expanding the term $-\frac{\tau^y}{\rho H^2}\frac{\partial H}{\partial x}$ (which is on the right hand side of the governing equation for ψ) in a series of the eigenfunctions $\phi_n(x)$. This is the same method that was mentioned in Chapter 2 for analysing unsteady but linear ocean dynamics.

Stepping back from all this mathematical detail, what we have is a wind stress forcing a train of continental shelf waves along the slope. The modes that are forced depend on the frequencies present in the wind stress forcing and are obtained by superposition, but once the waves have been forced the train propagates and must disperse. The original source for this model is Gill and Schumann (1974) and a few further details are reproduced in Gill (1982) (beware of the different definitions of x and y). In fact, the books by Gill (1982) and Leblond and Mysak (1978) are recommended for further general reading about continental shelf waves. In this text it has been decided to devote extra space later in this chapter to trapped waves around islands. The trapped nature of continental shelf waves should perhaps be emphasised here as it may have been missed. All the wave like characteristics are along the coast and away from the coast both the elevation and current simply decay. (Elevation is zero only in the ideal case, where there is a forcing wind, there is an elevation - it is simply dynamically unimportant for these waves). A little later Kelvin waves are introduced that have the same trapped property. Kelvin waves exist over a flat sea-bed; it is the sea surface elevation that is essential this time!

In reality, there is friction that dampen continental shelf waves, and the water is usually stratified at eastern edges of the ocean, witness the upwelling models of Chapter 3. Let us briefly describe models that take this into account. The kind of general modal decomposition model of the type outlined in Gill (1982), see Gill and Clarke (1974), cannot be used here as horizontal boundaries are difficult to incorporate. The alternative is to use continuous stratification as in the models of Brink (1991). Let us briefly examine these, the dynamics of which are governed by the linear system:

$$\frac{\partial u}{\partial t} - fv = -\frac{1}{\rho_0}\frac{\partial p}{\partial x} + \frac{1}{\rho_0}\frac{\partial \tau^x}{\partial z}$$

$$\frac{\partial v}{\partial t} + fu = -\frac{1}{\rho_0}\frac{\partial p}{\partial y} + \frac{1}{\rho_0}\frac{\partial \tau^y}{\partial z}$$

$$0 = -\frac{\partial p}{\partial z} - g\rho'$$

$$\frac{\partial u}{\partial x} + \frac{\partial v}{\partial y} + \frac{\partial w}{\partial z} = 0$$

$$\frac{\partial \rho'}{\partial t} + w\frac{\partial \bar{\rho}}{\partial z} = 0.$$

In this formulation which is Boussinesq, the density assumes the form

$$\rho(x,y,z,t) = \rho_0 + \bar{\rho}(z) + \rho'(x,y,z,t)$$

and

$$|\rho'| << \bar{\rho} << \rho_0.$$

The single equation for p is derived:

$$\frac{\partial}{\partial t}\left(\frac{\partial^2 p}{\partial x^2} + \frac{\partial^2 p}{\partial y^2}\right) + \left(f^2 + \frac{\partial^2}{\partial t^2}\right)\frac{\partial}{\partial z}\left(\frac{1}{N^2}\frac{\partial^2 p}{\partial z\partial t}\right) = 0$$

for the case where there is no wind stress. The buoyancy frequency is N where

$$N^2 = -\frac{g}{\rho_0}\frac{\partial \bar{\rho}}{\partial z}.$$

The equation for p is in fact the equation for vorticity for this stratified system. The complications introduced by stratification are all too obvious. In order to apply this equation to coastal trapped waves, it is assumed that variables (specifically p as all others can be expressed in terms of p) have wave-like behaviour:

$$p \sim \exp[i(\omega t + ly)]$$

but the equation for p even in only the two variables x and z has to be solved numerically. The methods used are based on a semi implicit scheme. The solutions represent trapped waves that are neither classical continental shelf waves (because of the stratification) but nor are they Kelvin waves (no free surface

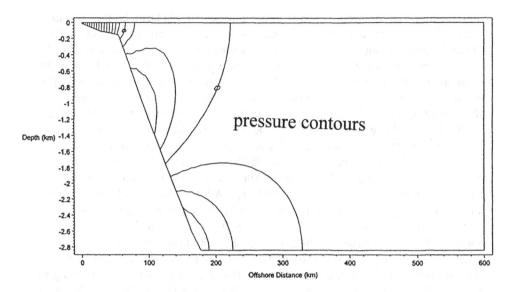

Figure 6.5: Stratified trapped waves

elevation). Figure 6.5 shows some of these waves and their trapping as represented by contours of pressure. In a stratified regime, the distinction between (baroclinic) Kelvin waves and continental shelf waves becomes artificial and not worth making. For forced waves where the single equation for p cannot be derived, the numerical procedure has to make use of an eigenfunction expansion as in the layered solution above, but this remains a research topic yet to be fully resolved.

6.5 Tides on Continental Shelves

One of the most distinctive features of continental shelf seas is the relative strength of the tidal currents compared with those that arise from other sources. From the point of view of fluid flow this feature is easy to explain. Just as flow through a large-diameter pipe accelerates to a faster value as the diameter decreases, so a slow tidal flow in the deep sea accelerates to fast tide over the shallow continental shelf. In shelf regions, tidal currents are usually ten times stronger than currents from other sources (wind or convection due to freshwater inputs from rivers). Let us begin this look at continental shelf tides.

At the outset, let us declare that we shall be concerned with practical tidal dynamics and not with the kind of theoretical tidal modelling which was published a hundred or more years ago, dating back to Laplace in 1776, involving periodic hydrodynamics on a rotating sphere.

Tides are due to the astronomical forces that arise from gravitational attraction between the Earth and the Moon and, to a lesser extent, the Earth and the Sun. These forces themselves act on the Earth's large bodies of water, the Pacific, Atlantic and Indian Oceans. Typically the moon pulls the surface of the Atlantic about half a metre from its mean level both when it is directly overhead and also when it is on the other side of the earth. This can be thought of as arising because the same gravitational attraction of the moon pulls the earth itself closer to the moon than the water which is furthest from the moon thus resulting in a second high water on the opposite side of the earth. These slight tides, and the equivalent depression when the moon is on either horizon causes a wave (two high waters a day as the earth rotates) which propagates. This wave passes over the continental shelf and, once on the continental shelf itself, increases in amplitude because of the conservation of mass (the same reason as the enhancement of tidal velocities). Another feature of tides is their wavelength. This is the horizontal distance between two successive high waters at any fixed time. This distance is typically 1000 km (for a tide with two high waters a day in a continental shelf region). The amplitude of a tide is but a few metres or even less. The wave slope, a common dimensionless parameter which is often used as a characteristic measure, is the ratio of these two quantities (see Figure 6.2) and is thus only about 10^{-5}. This means that tides, considered as water waves, are very long waves indeed. Some elements of shallow water wave theory were given earlier in this chapter. In Chapter 2, non-dimensional quantities were defined and as mentioned there, these certainly can apply to tides. In order to see how shallow these water waves are let us insert some typical values for the parameters. Since the depth of the continental shelf sea is only 200 m or so, and the ratio depth/wavelength is only 2×10^{-4}, we can consider the water as indeed very shallow. One consequence of this is the tidal currents should be virtually independent of the depth. This is certainly true if we are concerned with tidal height predictions. However, tidal currents are influenced by bottom friction due to the roughness of the sea bed; this can be of crucial importance. This was used as an example in Chapter 2 and we will return to it again later.

The tide can therefore be crudely modelled as a to and fro, depth- independent movement of the sea. Since the pressure is hydrostatic, that is only dependent on the weight of water above a particular point, the pressure is proportional to the elevation of the sea surface above mean sea level (taking the mean sea level as a reference level for pressure). Thus as mentioned in Chapter 2

$$p(x, y, z, t) = p_A + \rho g(\eta(x, y, t) - z).$$

From this, by differentiation

$$\frac{\partial p}{\partial x} = \rho g \frac{\partial \eta}{\partial x}$$

and
$$\frac{\partial p}{\partial y} = \rho g \frac{\partial \eta}{\partial y}.$$

Hence horizontal pressure gradients are proportional to the slope of the sea surface in a constant density sea. Even if the density depends on z this remains true, but in the case where $\rho = \rho(x, y)$ the thermal wind equations hold (see

Chapter 2, section 2.4.1). The dynamic balance required to model tidal motion in shallow water (continental shelf seas) is therefore a balance between particle acceleration due to the back and forth water movement of the tide together with Coriolis acceleration and the pressure gradient forces which manifest themselves as a sea surface slope. In a narrow inlet or estuary, tidal waves propagate up or down in more or less the same way as waves in a domestic bath can be driven by a convenient wiggle of a foot. However, tidal waves have a very long wavelength. This is as if the water in the bath rises and falls, keeping the surface horizontal (conservation of mass demands an infinitely long bath here!). All frictional effects have been ignored of course. The introduction of the Earth's rotation in the form of the Coriolis acceleration causes this wave to behave rather differently. It is as if the wave in the bath, instead of preserving a horizontal crest as high water comes toward you, is lopsided to the extent that the crest is higher to the left, falling away exponentially to the right. As it travels away from you, the highest point of the wave is on the right and the line of high water falls away exponentially to the left. Viewed from above the wave hugs the edge of the bath, propagating anticlockwise around it. Of course since the bath has a finite width and length, neither the exponential profile offshore nor the sinusoidal nature of the wave along the shore is perfect. Perfection only occurs with an infinitely long coastline next to an infinitely wide sea. This perfection is called a Kelvin wave, and although it is never found in pure form, the tides in many coastal seas e.g. the northern North Sea, are well approximated by them. Figure 6.6 shows a Kelvin wave together with a plan view in terms of cotidal lines and corange lines. The cotidal line is a line along which the maximum tidal amplitude is attained at the same time and it can be thought of as an instantaneous snapshot of high water taken at three hourly intervals. The corange lines are lines along which the range of the tide is the same. For a Kelvin wave, these lines form a square grid. Let us see how this square grid arises. The wave travels down the coast with its crest perpendicular to the coastline; hence cotidal lines, which must follow the crest line, are also perpendicular to the coast. On the other hand, along a line parallel to the coast the "tide" (that is, how much the sea surface rises and falls) will be the same. This is precisely the range, and hence corange lines are also parallel to the coast. Now compare this idealise square grid system of cotidal and corange lines with an actual picture of cotidal and corange lines taken from observations in the North Sea (Figure 6.7). You will see that, near the coast of the UK, in the north, the cotidal and corange lines actually form a square pattern. This tide is almost a perfect Kelvin wave, although perfection is absent elsewhere.

6.5.1 Modelling Kelvin waves

In order to be more precise, here briefly is a model for Kelvin waves. The basic balance is described above and symbolically is:

$$\frac{\partial u}{\partial t} - fv = -g\frac{\partial \eta}{\partial x},$$

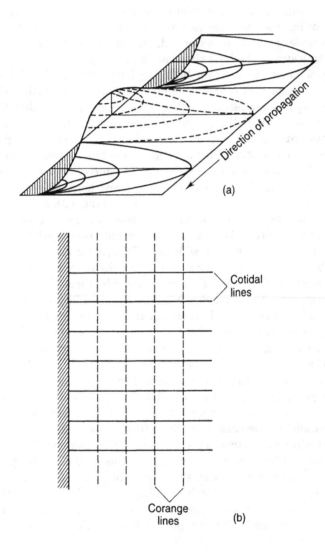

Figure 6.6: A Kelvin wave: (a) isometric view; (b) plan view

$$\frac{\partial v}{\partial t} + fu = -g\frac{\partial \eta}{\partial y},$$

$$\frac{\partial \eta}{\partial t} + h\left(\frac{\partial u}{\partial x} + \frac{\partial v}{\partial y}\right) = 0.$$

The first two equations are quasi-geostrophic balance where the horizontal pressure gradients have been put in terms of a sea surface slope via hydrostatic balance. The third equation is the conservation of mass in a sea of depth h and surface elevation $\eta(x,y,t)$ measured from mean sea level. The notation (u,v) for horizontal current f for Coriolis parameter and (x,y) for Cartesian axes x-East and y-North has been retained. These equations, first met in Chapter 2 are sometimes called the Laplace Tidal Equations although in this form they are due to Lord Kelvin. All non-linear terms and friction terms have been ignored. The idealised perfect Kelvin wave has an infinitely long coast on its right as it propagates. Thus we first assume that the wave is at a single frequency ω which means:

$$\frac{\partial \eta}{\partial d} = i\omega\eta, \quad \frac{\partial u}{\partial t} = i\omega u \text{ and } \frac{\partial v}{\partial t} = i\omega v.$$

Straight away therefore, we deduce that:

$$(\nabla^2 + \lambda^2)\eta = 0$$

where $\lambda^2 = \dfrac{\omega^2 - f^2}{gh}$. The same partial differential equation is satisfied by u and v. Moreover, the first two of the Laplace Tidal Equations become simultaneous equations in u,v and can be solved to give:

$$u = \frac{g}{\omega^2 - f^2}\left\{i\omega\frac{\partial \eta}{\partial x} + f\frac{\partial \eta}{\partial y}\right\}$$

$$v = \frac{g}{\omega^2 - f^2}\left\{-f\frac{\partial \eta}{\partial x} + i\omega\frac{\partial \eta}{\partial y}\right\}.$$

Now we are ready to impose the fundamental Kelvin wave assumption. There is no flow perpendicular to the coast, which with our co-ordinate system means that $u = 0$. There are many ways to extract the mathematical form of the Kelvin wave, my particular favourite is to note that if $u = 0$, then

$$i\omega\frac{\partial \eta}{\partial x} = -f\frac{\partial \eta}{\partial y}, \tag{6.9}$$

or

$$-\omega^2\frac{\partial^2 \eta}{\partial x^2} = f^2\frac{\partial^2 \eta}{\partial y^2}$$

upon differentiation. Since $(\nabla^2 + \lambda^2)\eta = 0$ which written out in full is:

$$\frac{\partial^2 \eta}{\partial x^2} + \frac{\partial^2 \eta}{\partial y^2} + \frac{\omega^2}{gh}\eta - \frac{f^2}{gh}\eta = 0,$$

the following two equations for η must hold

$$\frac{\partial^2 \eta}{\partial y^2} + \frac{\omega^2}{gh}\eta = 0$$

$$\frac{\partial^2 \eta}{\partial x^2} - \frac{f^2}{gh}\eta = 0.$$

In effect, demanding that $u = 0$ forces us to look for solutions of $(\nabla^2 + \lambda^2)\eta = 0$ in the form of pairing off the four terms to give one equation in x (with exponential solution) and one equation in y (with sinusoidal solution). Remembering that only the real parts have significance, we can thus write the Kelvin wave solution as:

$$\eta_1 = C_1 e^{-fx/\sqrt{gh}} \cos\left(\frac{\omega y}{\sqrt{gh}} + \omega t\right)$$

where there is a coast at $x = 0$ (y-axis) and C_1 is a constant.

If there is another coast at $x = b$ (the width of the Northern North Sea is b km.) then a Kelvin wave for that coast would be:

$$\eta_2 = C_2 e^{f(x-b)/\sqrt{gh}} \cos\left(\omega t - \frac{\omega y}{\sqrt{gh}}\right),$$

where C_2 is another constant. Writing $\eta_1 + \eta_2 = \eta$ gives a reasonable model for Kelvin waves in a canal. Physically, it is the direction of the Earth's rotation that makes the sea pile to the right as the Kelvin wave travels. Mathematically it is Equation 6.9 from which it can be seen that if $f < 0$ (southern hemisphere) then

$$\eta = C_3 e^{fx/\sqrt{gh}} \cos\left(-\frac{\omega y}{\sqrt{gh}} + \omega t\right)$$

(C_3 a third constant) is the correct solution and the sea piles to the left as the wave travels.

6.5.2   Numerical models of tides

In order to predict tidal elevations and the currents that arise from tides, it is usual to build a numerical model based on a grid designed to cover the desired basin, sea or coastal area as described in Chapter 3. Let us look at some of the processes it is essential to incorporate in a successful tidal model. Perhaps the best place to start is to see what can be ignored. First, it can be assumed that there is no weather! This rather startling assumption is put in context once it is realised that wind and pressure effects that arise from the weather can be added later. They are not essential ingredients of a tidal model, although everyone is aware of their importance to understanding flooding during storm surges (see section 6.8). The technical reason why we do this adding of effects is linked to the extremely long wavelength of tides (Figure 6.2). Tides are linear to the extent that meteorological effects can be added and there is very little dynamic interaction between wind and tide. Tides themselves, however, can exhibit highly non-linear characteristics. One only needs to think of shallow water where drying can occur, or peninsulas where a tidal current can change sharply in direction. Both of these scenarios have length scales within them that render the advective term (see Chapter 2) large enough to be important. When this term is important, the model is described as non-linear. However, in the modelling of tides there is no non-linearity connected with wind or pressure. (Contrast this with surface sea waves and wind which can and do interact and give rise to the phenomenon of Langmuir circulations; see section 6.7.1). Second, unless internal tides are of concern, it is reasonably safe to ignore changes in density. In coastal seas, density changes are normally due to freshwater input via rivers rather than to temperature contrasts, although summer thermoclines do exist. Again preliminary tidal modelling can be done accurately without involving these density fronts or pycnoclines. This is due principally to the difference in length scales associated with these density interfaces. These issues will also be addressed in section 6.7.1. If the prime concern is modelling tidal elevation, certainly the most important factor for shipping, then the model must include the following factors: particle acceleration (because tide is a wave, albeit a very long one, and the waves have a to and fro motion which implies acceleration); Coriolis acceleration (because the length scales of the waves is in general large enough for the rotation of the Earth to be important); sea surface slope (this slope is alone responsible for the horizontal pressure gradients that act as a horizontal force); and friction. This last term needs some explanation. Astronomical forces generate the tides, but as far as the seas of the continental shelf are concerned, they are driven by the oscillations of a larger neighbouring body of water (the ocean). What one has therefore is the standard set of partial differential equations, the Laplace Tidal Equations perhaps modified by the inclusion of non-linear terms and friction but discretised using finite differences. The friction is essential in a numerical model to dissipate the momentum at the bed, otherwise the solution to the equations will become too energetic - there is nowhere for it to go! At the open boundary, there is an incoming wave representing the oscillation of the neighbouring body of water. The actual scheme used is up to the user, but is usually semi-implicit as in that presented by Backhaus (1983) and outlined in Chapter 3, section 3.5.1. This is the classic numerical compromise between stability and truncation error. Tidal models are, perhaps surprisingly usually started from cold, that is a still region of sea is subjected to the open boundary oscillation and the incoming wave is allowed to propagate refract and reflect and radiate out until such time as the model settles down to a "steady state". One can perhaps now appreciate more the importance of the radiation condition at the open boundary. Without one that works well, models like this will not operate. As mentioned, the only mechanism available to dissipate the energy that is being so liberally pumped in via the continental shelf is friction, and this frictional dissipation has to take place at the sea bed. The form that friction takes has been outlined in Chapter 2. Normally, a quadratic law relating the drag to the square of the speed provides an adequate model that can also be justified on dimensional grounds.

If finite elements are chosen, see section 3.5.4 then the procedure of inte-

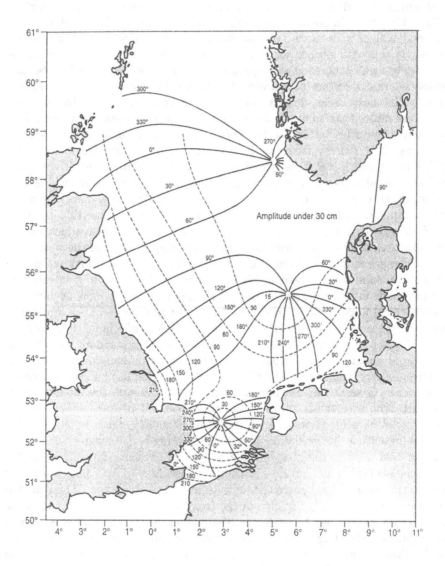

Figure 6.7: Cotidal and corange lines, denoted by full and dotted lines, respectively. The associated numbers give the values of γ (in degrees) and of H (in cm).

grating through each element and adding up contributions from all the elements means that the periodic forcing actually effectively acts as a body force. The act of integration has this effect; the limits of the integral are the boundary condition, and once the integration is performed, the variables evaluated at these limits are explicit in the equations.

Physically, the scenario is then rather like forced SHM where a system is forced by an oscillation and the response exhibits the same frequency with the addition of harmonics. These harmonics are due to non-linearities, either the advection terms from which M_4 tide stems from M_2 parent tide, or friction which can give rise to M_6. (M_2 is the semi-diurnal lunar tide). In a purely linear model, no harmonics can arise and the model will give the M_2 generated currents and elevations throughout the modelled region. Apart from very near the coast and where the sea is exceptionally shallow, linear tidal models with linear frictional dissipation at the sea bed are successful. Those that incorporate quadratic friction are even more so, so good in fact that they have not been surpassed by models with more elaborate sea bed friction. The case studies of Chapter 3 are good examples of actual numerical models although they all combine tide with wind as they represent an attempt to model real seas.

Figure 6.7 shows the accepted picture of cotidal and corange lines for the M_2 tide in the North Sea. Later in the chapter, Figure 6.13 shows the numerical simulation of this for the North Sea and surrounding area. The net picture is, loosely, a wave entering the North Sea keeping the coast on its right and exiting along the Norwegian coast. The wave is often very close to the theoretical Kelvin wave, especially along the Scottish and Northumbrian coasts. The wave gets distorted however as it travels around the North Sea, both by irregularities in the sea bed topography which cause refraction and by irregularities in the coast which cause reflection, and of course by frictional effects. A purely tidal model does predict the observed tide in the North Sea as can be seen by this figure. The numerical calculations were done on a grid of spacing $\frac{1}{3}^{\circ}$ latitude by $\frac{1}{2}^{\circ}$ longitude and the amphidromes are reproduced with reasonable accuracy. We return to this model of tides when considering storm surges in section 6.9. Wind driven currents are discussed next.

6.6 Wind Driven Currents

The other main cause of continental shelf sea currents is wind. The balance valid near the sea surface is one between the Coriolis acceleration and friction which causes Ekman circulation (see chapter 2, section 2.6.1). As we said there, these spiral currents are not easy to detect by measurements, since their structure is easily disturbed by local effects such as eddies and surface waves. The major influence of the wind on the ocean occurs, quite obviously, close to the surface of the sea. However, the dynamics of the ocean are complex and the fact that the sea is moving under the influence of the wind in turn drives deeper flows. In particular, there is a vertical velocity which results directly from the wind, and this vertical velocity when occurring at the base of the Ekman layer is termed

Ekman suction. A positive or negative Ekman suction will exert a strong effect on that part of the sea that is far below the surface and is not directly wind driven.

The Earth rotates, and a consequence of this rotation is that wind-driven flows do not move in the direction of the wind. In the Ekman layer, the net effect of a steady wind is a flow to the right of the wind (in the northern hemisphere; left in the southern hemisphere). Wind is never steady for long periods, so the Ekman currents are always being spun up or attenuating. Since the detailed structure of a pure Ekman current is spiral in nature (see Figure 6.6), even given the mythical steady wind, the real waxing and waning versions can bear a close resemblance to turbulent eddies. So far there has been no mention at all of coasts. Coastlines are a barrier to currents and all motion is forced to flow parallel to them. Modelling currents along coasts tends to be specific; there is very little to be said of a general nature, except that again the overall dynamic balance remains between frictional forces, Coriolis acceleration and pressure gradient forces. When there are bends in the coastline on a scale of hundreds of kilometres, then it is usually prudent to consider additionally the effects of advective acceleration. An effective way of assessing whether on not various terms should or should not be included in a particular model of a specific sea is to utilise dimensional analysis, as outlined in Chapter 2.

6.7   Estuarial Circulation

Another important feature of continental shelf seas are river mouths. The circulation in the saline mouths of many river outlets resembles that of a coastal sea. The Moray Firth and Firth of Forth in Scotland, and the mouths of the larger Norwegian Fjords are good examples. Further upstream, the presence of fresh water exerts a greater influence on the circulation. Salt water is heavier than fresh water. This gives rise to the 'saline wedge' picture often taken as typical of an estuary (see Figure 6.7). Modelling estuarial flows has a long tradition. The first approach, still of some use today, is to use box models (see section 5.3). In these models, the domain of the estuary is divided into conveniently sized boxes (either two-dimensional vertical or horizontal sections, or three-dimensional boxes), and an estimation is made of the transportation of various water properties across the boundaries of each box. These properties can include salinity, temperature, various chemical and biological constituents as well as flow. Once the boxes have their assigned parameter values, and each box contains water with properties that are consistent with the properties of the of the water in adjacent boxes and consistent with an overall agreed picture, then the model can be exploited. For example, an inflow at one point of the estuary will influence the water properties in one box or two, but will also probably influence the behaviour of the water in surrounding boxes through having to satisfy the conditions at the boundary of the boxes. These box models are still useful under two circumstances: first, if an approximate picture is required quickly, then a box model can give some insights; and second, if there is no

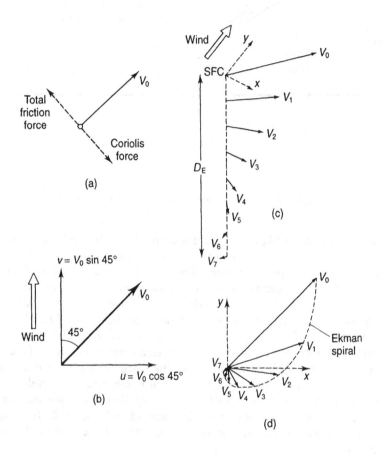

Figure 6.8: Structure of a pure Ekman current. (a) Forces and surface velocity; (b) plan view at surface; (c) perspective view; and (d) plan view, velocities at equal depth intervals, as in (c). From Pond and Pickard (1991). Reproduced with permission.

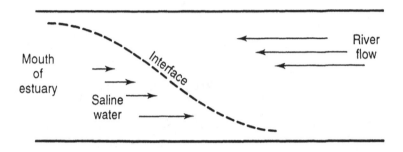

Figure 6.9: Saline wedge in a stratified estuary

appropriate software available to build a more elaborate model and, more importantly, only a sketchy knowledge in terms of observations in certain areas, then a box model can provide a rough first model. Box models can often be the initial models that trigger additional observation programmes or experiments. In an educational context too, box models can be a useful vehicle for introducing the rudiments of modelling. However, it must be said that if reliable predictions are the order of the day, then more sophisticated models that make use of software would normally be required. Some elementary dimensional analysis is also helpful. As estuaries are places where there are often density differences in the vertical, one might expect the Richardson number to be a guide to which processes to model. This is indeed true, except that there are various types of Richardson number in addition to that defined in Chapter 2. In Chapter 2 the Richardson number R_i was defined as the ratio of the square of buoyancy frequency to shear:

$$R_i = \frac{N^2}{|\partial u/\partial z|^2} = -\frac{1}{\rho_0}\frac{\partial \rho_0}{\partial z} \bigg/ \left|\frac{\partial u}{\partial z}\right|^2.$$

In an estuarial context, Fischer (1972) has defined the *estuarine* Richardson number, $R_i^{(e)}$ through the formula

$$R_i^{(e)} = \frac{\Delta\rho}{\rho}\frac{g u_r}{b\bar{u}^3}.$$

Here, u_r is the river flow averaged over the cross-section of the estuary, b is the breadth of the estuary and $\bar{u}$ is the root mean square tidal flow. We have used

$\Delta\rho$ to denote the change in density between sea water and fresh water with ρ being the density of the former. Another dimensionless quantity used more when one wishes to assess the stability of interfacial waves in the presence of currents is the densimetric Froude Number F_d defined by

$$F_d = \frac{u_r}{\sqrt{gh(\Delta\rho/\rho)}}.$$

This is clearly the ratio of river flow to the speed (celerity) of the interfacial or baroclinic wave. Where F_d changes from greater to less than one is often the definition of where the river ends and the estuary begins. The densimetric Froude number and the Estuarine Richardson number are the two dimensionless groupings that emerge from dimensional analysis. The direct application of the Buckingham Pi Theorem is tricky due to the presence of the already dimensionless ratio of the densities $\Delta\rho/\rho$, and the presence of several lengths and speeds.

The two principal reasons for modelling estuaries are firstly to understand the tidal propagation, its interaction with river flow and density currents in order to forecast sediment flows, coastal changes and other similar matters of interest to the general public. This is usually the province of the Civil Engineer. Secondly, the quality of the water is extremely important. It therefore falls to the modeller to answer important questions concerning likely concentrations of pollutants, movements of water borne material etc. which might arise due to adjacent industry, recreation or accident. In other words, the modeller plays an important role in the overall management of any given estuary. Normally, a one dimensional model is built using finite differences or finite elements which is then used to predict cross stream averages. On the other hand there are some theoretical models which can be mathematically quite demanding, see the works of Ron Smith, Smith (1980) in particular takes the above stratified estuarial model a good deal further and derives various estuarial regimes. At some locations longitudinal diffusion is important, in others it is the unsteady horizontal mixing that dominates. The important point is that Smith (1980) derives these with mathematical rigor. In the paper Smith (1979) some reasonably accessible results are derived, and these are now outlined. The problem is to assess the combined effects of buoyancy and diffusion so reference to Chapter 5 is first in order. There the advection-diffusion equation was, if not exactly "derived" let us say it was explained physically. It takes the form

$$\frac{\partial\bar{c}}{\partial t} + u\frac{\partial\bar{c}}{\partial x} = \frac{\partial}{\partial x}\left(\kappa_x\frac{\partial\bar{c}}{\partial x}\right) + \frac{\partial}{\partial y}\left(\kappa_y\frac{\partial\bar{c}}{\partial y}\right)$$

in one dimension. Here x will be along the centre line of the estuary, $\bar{c}$ is the vertically averaged concentration of the contaminant and κ_x and κ_y diffusion coefficients. Later in that chapter, successful models that followed particles were described. If axes are chosen that move with the centre of mass of the contaminant, then it is possible to derive a non-linear equation

$$\frac{\partial \bar{c}}{\partial t} = \frac{\partial}{\partial y} \left\{ \left[k_1 + \left(\frac{g\Delta\rho}{\rho} \frac{\partial \bar{c}}{\partial y} \right)^2 k_2 \right] \frac{\partial \bar{c}}{\partial y} \right\}.$$

In this equation, only spreading in the y (cross-stream) direction is considered and k_1 and k_2 are constants which experimental scientists relate to friction velocity etc. The expressions $k_1 = h\Delta\rho u^*/\rho$ and $k_2 = h^5 g^2/96 u^{*3} k^3$ have been used where h is the depth, k von Karman's constant and u^* a friction velocity. This equation is due to Erdogan and Chatwin (1967) and is referred to the Erdogan-Chatwin equation. What has happened here is the advection term (non-linear) has been swept up into a modified non-linear diffusion term which leads to a different, and often better description of the process. Solving this equation is however very difficult but in-roads have been made by Smith (1978). The ratio of eddy viscosity to eddy diffusivity (κ_y above) is dimensionless and referred to as the Schmidt number. The Erdogan-Chatwin description of diffusion and buoyancy was really only good for large Schmidt numbers. Various other formulations appropriate to estuaries are available, see the review Chatwin and Allen (1985). For a full appreciation of what balances are valid with various parameter ranges figure 5 of Smith (1979) or figure 5 of Smith (1980) should be consulted. No complicated mathematics will be done here.

One of the problems in using dimensional analysis here is hinted at above. As mass never features, it is only present in the dimensionless ratio $\Delta\rho/\rho$, what can be deduced is limited. Take for example the Erdogan-Chatwin equation just mentioned:

$$\frac{\partial \bar{c}}{\partial t} = \frac{\partial}{\partial y} \left\{ \left[k_1 + \left(\frac{g\Delta\rho}{\rho} \frac{\partial \bar{c}}{\partial y} \right)^2 k_2 \right] \frac{\partial \bar{c}}{\partial y} \right\}$$

and propose the scalings:

$$\frac{g\Delta\rho}{\rho} \sim \frac{U^2}{h} \left(\frac{u^*}{U} \right)^A, \text{ and cross stream length scale} \sim \frac{1}{h} \left(\frac{u^*}{U} \right)^B$$

where U is a typical estuarial flow speed, u^* is the friction velocity h the depth and A and B are constants to be determined. This is not strictly a dimensional analysis, but is an estimation of scales derived from it. If the above expressions are substituted into the Erdogan-Chatwin equation, then the two terms on the right are of the same order if $A = -B$. Smith (1979) gives arguments to support putting $A = 1$ but only if the downstream flow term which is absent from the Erdogan-Chatwin equation is also included. Note also that it is assumed that the ratio u^*/U is a small parameter. It turns out that the above scaling returns us to linear diffusion, neglecting the interaction terms. The buoyancy acceleration is proportional to u^*U and cross-stream gradients proportional to u^*/U. The modelling assumption that the friction and buoyancy are of the same order is all too apparent here. Let us leave estuarial models now and turn to something different.

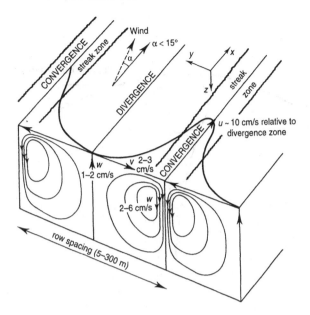

Figure 6.10: Langmuir circulation

There is an interesting phenomenon not restricted to estuaries but of a similar scale and certainly observed there called Langmuir circulation. Langmuir circulation is the name given to vortices that occur near the surface of the sea. A few years ago, it could be said that not everyone agreed about their existence, but now it is accepted that they are always present when the wind is stronger than 7 metres per second. What are these vortices? They are near-surface circulation that arises as a result of the interaction between the waves on the surface of the sea and wind-driven flow. In typical Langmuir circulation (see Figure 6.10) the fluid moves in a path reminiscent of a coiled spring, spiraling with the axis of the spring horizontal and more or less parallel to the wind. The vortices have a spacing of about 200 m in the sea, although analogous circulation is found in lakes and here the spacing is much less, in fact only in the order of tens of metres. Detailed modelling of Langmuir circulation is not very easy since very different mechanisms need to be incorporated to include both wind-driven currents and currents due to waves. When modelling waves, the first assumption is that the profile of the waves themselves can be approximated by a sinusoidal shape under which the water particles follow closed circular (or nearly circular) paths; see Figure 6.11. Unfortunately, such a model of waves cannot predict any net movement of water or indeed waterborne material. If this is surprising, look at a cork bobbing on a wavy sea; it may have a small net drift, but the principal movement is circular. Also look again at section 6.3 where linear wave theory is briefly introduced. This is the same for all water particles. The assumption of a purely sinusoidal surface wave in fact eradicates any wave drift - there is no wave drift under a sinusoidal sea. It is thus essential

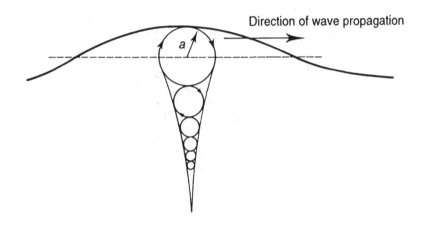

Figure 6.11: Particle motion under a deep water wave: a is the wave amplitude

to include the non-linear or advective terms in any model that is to predict transport of material. It is this combination of a non-linear wave theory, which can predict a drift (the wave drift or Stokes' drift) see section 6.7.1, with a direct wind drift that can successfully model Langmuir circulation. The dimensionless number associated with wave theories is the wave slope which is the ratio of the wave amplitude to the wave length. An appropriate dimensionless number associated with pure wind drift is the vertical Ekman number for flows whose vertical scale is typically hundreds of metres. Such a scale is, however, usually inappropriate. The only alternative is to examine the pure wind-driven current in terms of the direct action of the air on the sea. This can be done through the stress-rate of strain relationships of the type used to define eddy viscosity (see Chapter 2). We then look at the ratio of this wind drift to the pure wave or Stokes' drift. If this number is around unity, then we can expect both of these effects to be equally important. A model that includes all the relevant non-linear wave and wind drift terms is then built, and Langmuir circulation is predicted. Another approach is to use the Froude number, which denoted the relative importance of current (in this case wind-driven flow) and wave speed. This then leads to a stability criterion for Langmuir circulations and takes us outside the scope of this text. Perhaps the only good news is that the scale is such that the Coriolis terms can be safely ignored since the typical length scale, even of the largest Langmuir cells, is still deemed small when compared with the length scale associated with the Coriolis acceleration (this is Rossby radius of deformation, U/f, where U is a typical speed and F is the Coriolis parameter, which is of the order of hundreds of kilometres; see section 2.5).

6.7.1   Models of Langmuir Circulation

There are no simple models of Langmuir circulation. The passage at the end of the last paragraph is an attempt at describing what is going on physically. We give here a basic model due to Craik and Leibovich (1976). The dominant motion is assumed to be surface (gravity waves) which have an amplitude which is small compared with depth. Friction is in the form of a constant eddy viscosity ν_v, and it is necessary to expand variables in terms of the (small) wave slope parameter ε, so

$$\mathbf{u} = \varepsilon \mathbf{u_0} + \varepsilon^2 \mathbf{u_1}$$

where $\mathbf{u_0}$ is the velocity due to linear water waves over infinite depth and $\mathbf{u_1}$ is the second order correction. The linear water wave equations thus give rise to a vorticity balance:

$$\frac{\partial \omega}{\partial t} = \nabla \times (\mathbf{u} \times \omega) + \nu_v \nabla^2 \omega$$

where $\omega = \nabla \times \mathbf{u}$ is the vorticity vector. Since $\mathbf{u_0} = \nabla \phi$ with $\nabla^2 \phi = 0$ which encapsulates linear water wave theory, systematic expansion in powers of ε is possible. Doing this, we have:

$$\omega = \varepsilon^2 \nabla \times \mathbf{u_1}$$
$$\mathbf{u_1} = \mathbf{u_{10}} + \varepsilon \mathbf{u_{11}}$$
$$\nabla \times \mathbf{u_1} = \omega_0 + \varepsilon \omega_1 + \dots$$

This enables $\mathbf{u_{10}}$, $\mathbf{u_{11}}$, ω_0 and ω_1 to be separated into mean and fluctuating parts. The equations are then followed through with time averaging taking place. The essential point being that it is the non-linear term $\nabla \times (\mathbf{u} \times \omega)$ which fuels all the subsequent interaction and is the source of the corkscrew motion characteristic of Langmuir Circulation. Later papers provide evidence of a connection between this theory and a more comprehensive theory of the Lagrangian mean (too advanced for this text, but see section 6.8.1).

6.8   Waves and Flows around Islands

In this section, we shall look at the modelling of waves and then flows which are trapped around islands. The history of modelling the diffraction of waves around islands really started in the civil engineering community with the modelling of the diffraction of waves by cylindrical piles. The cylindrical geometry means a different co-ordinate system, but the mathematics is reasonably easy to solve if the depth of the sea is large compared with the diameter of the cylinder. The model is a little trickier if this is not the case, but exact solution is still possible. As we want to discuss both the consequences and some extensions of this model, we shall give some technical details. The factors that are important to consider are the variations in depth which leads to the refraction of waves and can ultimately, singlehandedly as it were, lead to trapping. Those familiar with Snell's Law will understand how refraction can entrap waves. Then there is the Coriolis acceleration. This leads to the distortion of waves and the prevention of standing waves. If the Coriolis acceleration is large enough, the wave like character radially away from the island can disappear completely and what remains is a Kelvin like wave running around the island clockwise (in the Northern Hemisphere). This is also a trapped wave and occurs even if the depth is constant. Therefore there are two mechanisms that aid the trapping of waves, and wave trapping is certainly important as it leads to increased amplitudes as well as the amplification of previously small offshore signals. What follows is a brief look at models that explain the trapping of waves and their associated currents. Those interested in a more extensive treatment are directed towards the texts by Leblond and Mysak (1978) and Dingemans (1997).

We are concerned with islands, therefore it is natural to use plane polar (r, θ) co-ordinates. The shallow water equations are

$$\frac{\partial u}{\partial t} - fv = -g\frac{\partial \eta}{\partial r}$$

$$\frac{\partial v}{\partial t} + fu = -\frac{g}{r}\frac{\partial \eta}{\partial \theta}$$

$$\frac{\partial \eta}{\partial t} + \frac{1}{r}\left\{\frac{\partial}{\partial r}(hru) + \frac{\partial}{\partial \theta}(hv)\right\} = 0.$$

The non-linear terms are assumed negligible and there is no friction. If the time dependence takes the form of a wave of single frequency ω perhaps a tide, then

$$\eta = A(r, \theta)e^{i\omega t}$$

and we have that

$$\frac{\partial \eta}{\partial t} = i\omega\eta.$$

In plane polar co-ordinates (r, θ) we have

$$\nabla^2 \eta = \frac{\partial^2 \eta}{\partial r^2} + \frac{1}{r}\frac{\partial \eta}{\partial r} + \frac{1}{r^2}\frac{\partial^2 \eta}{\partial \theta^2}$$

and the equation for η turns out to be the rather lengthy

$$\nabla^2 \eta + \frac{\omega^2 - f^2}{gh}\eta + \frac{1}{h}\frac{\partial h}{\partial r}\frac{\partial \eta}{\partial r} + \frac{1}{r^2 h}\frac{\partial h}{\partial \theta}\frac{\partial \eta}{\partial \theta} + \frac{if}{\omega h r}\left[\frac{\partial h}{\partial \theta}\frac{\partial \eta}{\partial r} - \frac{\partial h}{\partial r}\frac{\partial \eta}{\partial \theta}\right] = 0.$$

The flow that results from this wave is quasi-geostrophic, that is basically geostrophic (balance between pressure gradient force and force resulting from Coriolis acceleration) but also slowly varying. Since the time dependence does only consist of oscillation at a single frequency we can easily derive the following equations for radial current u and transverse current v:

$$u = \frac{g}{\omega^2 - f^2}\left[\frac{f}{r}\frac{\partial \eta}{\partial \theta} + i\omega\frac{\partial \eta}{\partial r}\right]$$

$$v = \frac{g}{\omega^2 - f^2}\left[\frac{i\omega}{r}\frac{\partial \eta}{\partial \theta} - f\frac{\partial \eta}{\partial r}\right].$$

Some of you might still be worried about the presence of the imaginary unit $i = \sqrt{-1}$ in these expressions. All this means is that there are phase differences between $\eta(r, \theta)$, $u(r, \theta)$ and $v(r, \theta)$. The presence of rotation has this effect. If we set $f = 0$ in the long equation for η we can re-derive an equation that can yield solutions of interest to civil engineers which is applicable to waves diffracted around small islands. The smallness being defined by the radius being much less than the local Rossby radius of deformation but is also related to the frequency ω being small enough to compare with local value of the Coriolis parameter f. If the diameter of the island is a, then the Rossby number is U/fa where U is a typical current speed. The Rossby radius of deformation is $\sqrt{gh}/f$, and in order to ignore f we require:

$$a << \sqrt{gh}/f \quad \text{or} \quad af << \sqrt{gh}.$$

Thus we have that although U/fa is negligibly small if the denominator (fa) is replaced by the much larger $\sqrt{gh}$, the Froude number $U/\sqrt{gh}$ is certainly very small. Conversely, even though the Froude number is small (waves more important than currents) it may not be the case that the non-linear terms can be ignored. The current associated with a wave is of magnitude $Ag/\omega\lambda$ where A is the amplitude of the wave and λ its wavelength. The Froude number is thus $Ag/\omega\lambda\sqrt{gh}$ and its smallness is another way of saying that, after using the dispersion relation for $h << \lambda$, that $A << h$ or the amplitude is much lass than the water depth. This is the classic small amplitude wave assumption.

In order to progress, it is also assumed that the depth profile is axisymmetric of the form $h(r)$ so that all terms involving the derivatives of h with respect to θ are zero. The solution can now be obtained in terms of what is a standard technique of separating the variables:

$$\eta(r, \theta) = Z(r)\Theta(\theta).$$

Note that even though there is no variation of depth with tangential co-ordinate θ, it certainly cannot be assumed that the surface elevation η is also independent of θ. The separation technique yields the two equations for $Z(r)$ and $\Theta(\theta)$:

$$\frac{d^2Z}{dr^2} + \frac{1}{r}\frac{dZ}{dr} + \frac{1}{h}\frac{dh}{dr}\frac{dZ}{dr} + \left(\frac{\omega}{gh} - \frac{n^2}{r^2}\right)Z = 0$$

and

$$\frac{d^2\Theta}{d\theta^2} = -n^2\Theta$$

where the separation constant n is an integer. The negative sign in the second of these equations ensures that the solution $\Theta(\theta)$ is sinusoidal and n is an integer because there must be an exact number of waves around the island. The equation for $Z(r)$ is more interesting. Writing it in self adjoint form gives

$$\frac{d}{dr}\left[rh\frac{dZ}{dr}\right] + r\left[\frac{\omega^2}{g} - \frac{n^2}{r^2}h\right]Z = 0.$$

This enables us to deduce that as long as the inequality

$$\frac{\omega^2}{g} > \frac{n^2 h}{r^2}$$

holds then the positive nature of the coefficient of Z indicates that the solution $Z(r)$ will assume an oscillatory character as r increases. On the other hand if

$$\frac{\omega^2}{g} < \frac{n^2 h}{r^2}$$

$Z(r)$ will behave exponentially as r increases. The increasing term is rejected on physical grounds, therefore $Z(r)$ will in this case decay to zero as r increases. We call this *wave trapping* and trapping is thus assured as long as $h(r)$ is a function that grows faster than r^2. In this case a value of r will eventually be reached for which the solution decays exponentially. The zero mode ($n = 0$) is special and there will never be complete trapping in this case. From a practical point of view, $h(r)$ is never going to increase as rapidly as r^2 except perhaps very locally. In fact, over a large distance the depth will vary stochastically about the mean depth of the ocean thus ensuring $h(r)$ is in fact bounded for large r. Perfect trapping thus does not occur in practice. In order to say something about the flow, we note that for the case of no Coriolis acceleration,

$$u = \frac{ig}{\omega} \frac{\partial \eta}{\partial r}$$

$$v = -\frac{ng}{r\omega} \eta.$$

If $n = 0$ (zero mode) then $v = 0$ and all flow is in the direction radially away or towards the centre of the island. Note that this is the "ripples due to a stone thrown into a pond" solution. The elevation is 180° out of phase with the (radial) current as common to all one dimensional waves. The generally varying $h(r)$ case is still not possible to solve analytically, but if it assumed that h is a constant then it is possible to express the elevation η as follows:

$$\eta(r, \theta) = (AJ_n(kr) + BY_n(kr))e^{in\theta - i\omega t}$$

where $k = \dfrac{\omega}{\sqrt{gh}}$ and J_n, Y_n are the Bessel functions of the first and second kind respectively. (See the tome by Watson (1922) for all there is to know about Bessel functions. All you actually need to know here is that they are mathematical functions that oscillate and decay as $r \to \infty$ and as this happens, the period gets larger. Much like a vertical cross section, $\theta = $ constant, through the ripple in a pond due to a thrown stone mentioned above in fact!) It is tempting and possibly advantageous at this stage to discuss sea mounts. As these a do not break the surface, the wave must be finite at $r = 0$ whence the $Y_n(kr)$ term must be zero. The mathematics is thus simpler; simpler too because there are no island boundary conditions to consider. The temptation will however, be resisted! (see Leblond and Mysak (1978)). The advantage of

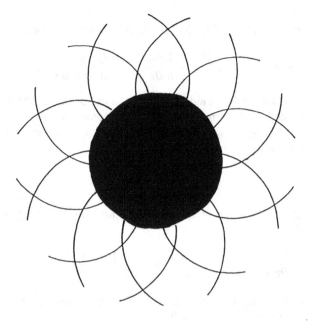

Figure 6.12: The spiraling waves

possessing exact mathematical solutions is that precise deductions are possible. For example, the solution for large n (high mode number) is virtually trapped. This is because once outside a particular critical radius, $k = \dfrac{n\sqrt{gh}}{\omega}$ the solution exhibits asymptotic behaviour which is that of a decaying exponential. The two sets of waves:

$$\eta(r,\theta) = J_n(kr)e^{(in\theta - i\omega t)}$$

and

$$\eta(r,\theta) = Y_n(kr)e^{(in\theta - i\omega t)}$$

represent two sets of spiraling waves that are not refracted (no depth change) and that propagate away from the island. The flow due to this type of wave can be readily deduced since

$$u = \frac{ig}{\omega}\frac{\partial \eta}{\partial r} \qquad \text{and} \qquad v = -\frac{ng}{r\omega}\eta$$

and, for large r this becomes predominantly radial as both $\dfrac{\partial \eta}{\partial r}$ and η have an $r^{-1/2}$ dependence asymptotically. The steady streaming (Stokes' drift) due to sinusoidal waves of this type is to be considered later in this chapter.

So far, nothing has been said about the effects of the earth's rotation. To see how their inclusion might effect things, let us take advantage of the earlier derivation of the general equation for the elevation $\eta(r,\theta)$ and keep a non-zero Coriolis parameter f but still make the assumption $h = h(r)$, that is the depth contours are axisymmetric. The (only slightly) simplified equation for η is

$$\nabla^2 \eta + \frac{\omega^2 - f^2}{gh}\eta + \frac{1}{h}\frac{\partial h}{\partial r}\frac{\partial \eta}{\partial r} - \frac{if}{\omega hr}\frac{\partial h}{\partial r}\frac{\partial \eta}{\partial \theta} = 0.$$

The last term on the left hand side prevents the use of the method of separation of variables, which is unfortunate. Nevertheless some progress can be made. As we are still considering the waves around an island, the variation of $\eta(r, \theta)$ must be periodic in θ, hence we can still write:

$$\frac{\partial \eta}{\partial \theta} = in\eta.$$

This enables the elimination of the variable θ and the following equation for η can be derived.

$$\frac{\partial}{\partial r}\left[rh\frac{\partial \eta}{\partial r}\right] + r\left[\frac{nfh'}{\omega h} + \frac{\omega^2 - f^2}{g} - \frac{n^2}{r^2}h\right]\eta = 0.$$

We have written h' for the derivative of h with respect to r. For trapping therefore the criterion is the inequality

$$\frac{nfh'}{\omega} + \frac{\omega^2 - f^2}{g} < \frac{n^2}{r^2}h.$$

So if $n = 0$, the zero mode, we can still get trapping provided $\omega < f$ (the Kelvin like wave mentioned earlier). As in practice we have

$$\frac{h}{r^2} \to 0 \qquad \text{as} \qquad r \to \infty$$

the above inequality in effect reduces to

$$\frac{nfh'}{\omega} + \frac{\omega^2}{g} < \frac{f^2}{g}.$$

We can thus say that trapping may still be prevented by large positive gradients in the sea bed with respect to r. However as $h' \to 0$ as $r \to \infty$ trapping is virtually assured for waves of subinertial frequency ($\omega < f$). This trapped wave is not a true Kelvin wave of course because the curvature of the coast prevents the precise decoupling between oscillatory behaviour (along the coast) and exponential behaviour (perpendicular to the coast) which are the essential feature of Kelvin waves. Nevertheless, they are trapped no matter how h varies with r which is very different from the situation with $f = 0$. When $f < \omega$, it seems there is unlikely to be trapping. The presence of stratification has and will not be considered in any detail here. Suffice it to say that a modal separation indicates that trapping is even more unlikely; a deduction confirmed by the recent numerical experiments of Brink (1999) who used continuous stratification. Finally, numerical (and some analytical) modelling on the interaction of waves and flows with real, not axisymmetric islands has been done. It is not difficult to deduce (see Leblond and Mysak (1978)) that asymmetry leads to the destruction of the standing nature of the waves. The lowest order modes exhibit a 180° phase difference between points on opposite sides of the island. Analytical models using elliptical islands have been used to represent tides around the Hawaiian Island of Oahu (see Reynolds (1978)) as well as Macquarie Island in Australia (Summerfield (1969)). Measurements on Oahu, on Cook Reef (north west of Australia) and Bermuda in the West Indies all indicate that some wave trapping is possible and that such a wave propagates around the island. Phase differences have also been measured in approximate agreement with models. Of course, validating models is rendered difficult as there is considerable noise so how do you tell whether or not there is a "perfectly" trapped wave? Other reasons for modelling such waves however have emerged. Signals from internal waves propagating from Cook Reef (for example) prove important indicators of seismic activity. This could help in the fast detection of underwater earthquakes. Similar indicators could also help us to understand the migration of certain marine organisms and build a better picture of the ecology of the area.

If the island is large enough, it is possible that the change of Coriolis parameter with latitude will be important. In this case, Rossby waves can be generated as waves diffract or as a large current impacts upon an island. In this case, even for wave frequencies that are well below the local value of the Coriolis parameter there is no trapping as the Rossby waves leak their energy westward. Such large scale flows are not really of concern here and interested readers are directed towards Leblond and Mysak (1978). They feature in the behaviour of tsunamis. Wave trapping off islands and continental shelves is still an active research area.

6.8.1 Steady flows

Flows due to waves will of course be primarily oscillatory in character. However, even perfectly sinusoidal waves will give rise to a steady current albeit a slow one. The reasons for this are not obvious, but the non-linear nature of fluid dynamics lies behind it. It was the nineteenth century mathematician and fluid dynamicist G G Stokes (1819 - 1903) who first showed that sinusoidal surface water waves must have a steady drift and the primary (but not the only) drift is named after him. The two ways of modelling the motion of a fluid are called Lagrangian and Eulerian. The Lagrangian model is where the observer is attached to a particular fluid particle and rides around on it (see section 5.4.2) . The Eulerian model is the standard one in which there is a fixed origin and axes, and the fluid is modelled relative to these. If $\mathbf{U}_L$ is the Lagrangian flow and $\mathbf{U}_E$ the Eulerian flow then:

$$\mathbf{U}_L = \mathbf{U}_E + \mathbf{U}_S$$

where $\mathbf{U}_S$ is the Stokes' drift. The account of Longuet-Higgins (1953) is the classic reference for the principal steady flows brought about by water waves. His detailed derivation is highly mathematical and requires looking at marked particles and tracking them as in the modelling of diffusion (see chapter 5), but the primary steady current due solely to the interaction of first order terms is the Stokes' drift defined by

$$\mathbf{U}_S = \overline{\left(\int \mathbf{u} dt . \boldsymbol{\nabla} \right) \mathbf{u}}.$$

In this expression, $\mathbf{u}$ must be purely oscillatory and the overbar represents the time average over a period of the oscillation. Often, the average over a few periods is taken when estimating the Stokes' drift in the field. Before investigating the Stokes' drift in detail, it is worth noting that there have been a number of interesting developments recently regarding the Lagrangian mean flow over the years, most of them stemming from the application of the conservation of potential vorticity (see section 2.4.3) and incorporating the change of Coriolis parameter with latitude. If the oscillation is due to a single frequency, then it can be shown (Moore (1970)) that if there are no closed contours, or closed *geostrophic* contours if f varies, then $\mathbf{U}_L$ is zero throughout. This means that $\mathbf{U}_E$ is precisely equal and opposite to the Stokes' drift $\mathbf{U}_S$.

Around and island there *are* closed contours and therefore one would expect $\mathbf{U}_L$ to follow the depth contours. In the calculations performed here, $\mathbf{U}_E$ (and therefore $\mathbf{U}_L$) have not been considered, but the calculation of $\mathbf{U}_S$ is straightforward as it is calculated solely from the primary flow. Of course if h is a constant Moore's theory tells us nothing at all about $\mathbf{U}_L$ other than it is arbitrary! For the situation here with non zero but constant f, $h = h(r)$ and

$$\eta(r, \theta, t) = C(r) e^{i(n\theta - \omega t)},$$

then the calculation of the Stokes' drift velocity is possible explicitly. Most of the steps in the calculation are omitted, but

$$\begin{aligned} \mathbf{U}_S &= \overline{\left(\int \mathbf{u} dt . \boldsymbol{\nabla} \right) \mathbf{u}} \\ &= \overline{\left(\frac{1}{i\omega} \mathbf{u} . \boldsymbol{\nabla} \right) \mathbf{u}} \\ &= \frac{g^2}{(\omega^2 - f^2)^2} (U_S, V_S) \end{aligned}$$

where the components of the Stokes' drift (U_S, V_S) are given by the rather lengthy expressions:

$$U_S = \overline{-i\omega\eta_r \left(\frac{n^2}{r^2}\eta - \eta_{rr} \right) - \frac{in^2 f^2}{\omega r^3}\eta^2 - \frac{in^3 f}{r^3}\eta^2 - \frac{inf}{r^2}\eta\eta_r + \frac{inf}{r}\eta\eta_{rr}}$$

and

$$V_S = \overline{\frac{n^3\omega}{r^3}\eta^2 - \frac{n\omega}{r}\eta_r^2 + \frac{n\omega}{r^2}\eta\eta_r + \frac{n^2 f}{r^3}\eta^2 - \frac{nf^2}{\omega r}\eta\eta_{rr} - f\eta_r\eta_{rr} + \frac{n^2 f}{r^2}\eta\eta_r + \frac{nf^2}{\omega r}\eta_r^2,}$$

where only the real parts have any physical significance and we have (just here and nowhere else in this text) used the suffix derivative notation whereby

$$\frac{\partial \eta}{\partial r} = \eta_r, \quad \frac{\partial^2 \eta}{\partial r^2} = \eta_{rr} \text{ etc..}$$

We thus see immediately that the radial component of $\mathbf{U}_S$ is zero (because all terms are imaginary). Without the Coriolis parameter we have the simpler expression

$$\mathbf{U}_S = \frac{g^2}{\omega^3} \left\{ \overline{i\eta_r \left(\frac{n^2}{r^2}\eta - \eta_{rr} \right)}, \overline{\frac{n^3}{r^3}\eta^2 - \frac{n}{r}\eta_r^2 + \frac{n}{r^2}\eta\eta_r} \right\},$$

also of course with zero radial component. Note that $\mathbf{U}_S = \mathbf{0}$ for the mode $n = 0$ which must be the case from symmetry. There is a useful expression for calculating the average of products of the real parts of quantities such as we have here, and this is

$$\overline{\Re\{z_1\}\Re\{z_2\}} = \Re\{\frac{1}{2}z_1 z_2^*\} = \Re\{\frac{1}{2}z_1^* z_2\}$$

where * represents complex conjugate. For the skeptics, here is a proof:
Let $z_1 = Ae^{i(n\theta - \omega t)}$ where $A = A_1 + iA_2$ so that

$$\Re\{z_1\} = A_1 \cos(n\theta - \omega t) - A_2 \sin(n\theta - \omega t).$$

Let $z_2 = Be^{i(n\theta - \omega t)}$ where $B = B_1 + iB_2$ so that

$$\Re\{z_2\} = B_1 \cos(n\theta - \omega t) - B_2 \sin(n\theta - \omega t)$$

then

$$\overline{\Re\{z_1\}\Re\{z_2\}} = \frac{1}{2}(A_1 B_1 + A_2 B_2) = \Re\{\frac{1}{2}z_1 z_2^*\} = \Re\{\frac{1}{2}z_1^* z_2\}$$

as required. It is possible to deduce similar expressions valid in a two layer sea, but more general extensions to deducing flows in stratified seas are the subject of present research. The work of Brink (1999) is mainly concerned with wave trapping, but the calculation of currents is certainly possible from his numerical models, and it is to numerical models one must look next for progress. For super-inertial frequencies ($\omega > f$) the work of Dale and Sherwin (1996) needs to be adapted to islands, which has not yet been done. One suspects trapping only at very special frequencies, and what currents are induced have yet to be revealed. This then represents a very fertile area for future research; some of you may be tempted to take it up.

6.9 Storm surge modelling

In this section we shall see how to put modelling to the very practical use of forecasting storm surges. The tides of the North Sea are largely M_2 that is they are due to the earth revolving under the moon. There are two high waters a day. The accepted pattern of cotidal and corange lines is shown in Figure 6.7. This pattern was established through measurement in the 1920s. The three points where the tide vanishes are called *amphidromic points*. The tide along the Northumbrian coast is very close to a perfect Kelvin wave, but distortion is introduced through the variation in the depth and the deviation of the coastline from straight. Frictional effects are also important and these are also absent from the pure Kelvin wave of course. A purely tidal model of the North Sea tides based on a uniform grid of spacing $\frac{1}{3}^{\circ}$ latitude and $\frac{1}{2}^{\circ}$ longitude gives the cotidal and corange lines of Figure 6.13. This model uses a semi-implicit finite difference scheme whereby a mesh is overlaid on the North Sea and variables computed according to an Arakawa B grid (see chapter 3). Later models used the C grid. The model is essentially two dimensional as the surface elevation is of prime concern and not how the currents vary with depth. The predicted pattern of cotidal and corange lines in the North Sea is certainly acceptably accurate. In these tidal models, there must be a friction term in order to extract the energy of the tidal forcing that is continually being fed in at the open boundary. This friction usually takes the form of a quadratic law which we have met as an example of dimensional analysis. Alternatively, for three dimensional models one could use eddy viscosity, but a vertical integration can still preserve depth dependent information by the careful use of transformations as done by Heaps (1971). Even with these sophisticated additions, a purely tidal model is of limited practical use, although an accurate tidal model is certainly an essential starting point. The one used to produce this figure is accurate in both amplitude and phase to 10%. In order to include meteorological effects, stress terms are fed in at the sea surface. The surface elevation predicted by the output of such models is therefore due to both tide and meteorology (pressure and wind). In order for the artificial open boundary at the shelf edge to have minimal effect, a *radiation condition* is imposed that ideally allows waves to exit freely from the domain of the numerical model with zero reflection. Any reflection at the open boundary would eventually render the solution over the entire domain inaccurate. Fortunately the friction in the system prevents this, but effects can be minimalised by careful application of an optimal radiation condition. Finding such an optimal condition is difficult and still taxing the brains of numerical analysts. One of the first successful three-dimensional models of the North Sea was built by Norman Heaps in the mid 1960s for the specific purpose of predicting storm surges from the knowledge of current tide, wind and pressure. With later developments this model of storm surges in the North Sea can be considered successful and is used operationally. For example in decisions as whether or not to employ the Thames barrier. The practicalities of storm surge prediction is summarised in the flow chart of Figure 6.14. This particular chart is due to Norman Heaps (1987). In the last thirty years or so there have been modifications to the storm surge model, principally by Roger Flather and his colleagues at the Proudman Oceanographic Laboratory. Present day models include a representation of non-linear processes. There was an attempt to use statistical hindcasting with a view to longer term prediction, but this proved

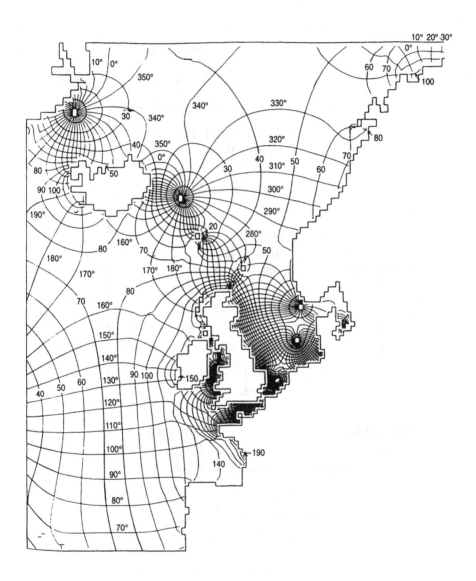

Figure 6.13: The output from a numerical tidal model for the north-west European continental shelf.

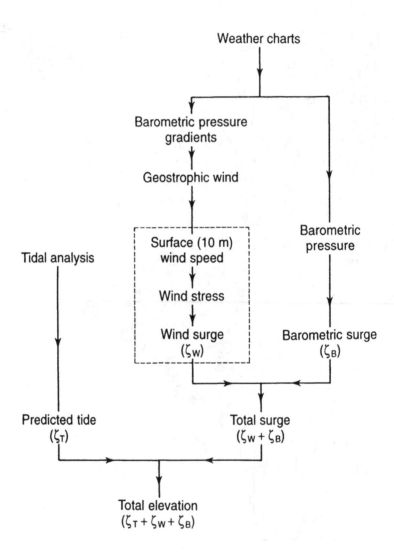

Figure 6.14: A storm surge flow chart.

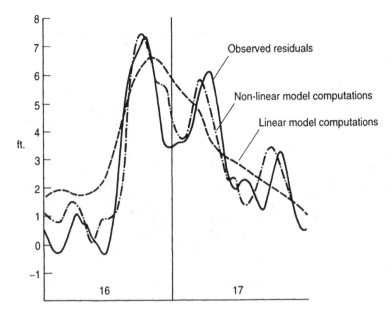

Figure 6.15: Wind induced surge levels at Southend, 16-17 February 1962. From Banks (1974). Reproduced with permission.

unreliable and is no longer operative. When the meteorological situation and the state of the tide (spring tide) combine in such a way as to make a storm surge even a remote possibility, the storm surge warning system swings into action. The UK Meteorological Office computer produces weather forecasts routinely and these are 24 hour, 48 hour and 72 hour predictions of the synoptic weather pattern over the UK. If the weather pattern seems likely to develop favourably, i.e. a low pressure centre deepening and becoming slow moving with its centre somewhere between the Shetland Islands and Norway, then the storm surge model is run. The spring tide high water is modelled as it travels down the UK coast as a distorted Kelvin wave. It takes about 12 hours to move from Orkney to the Thames so there is reasonable notice for any action which might include operating the Thames barrier, distributing sandbags, issuing various warnings to yachtsmen via coastguards etc. As the first model runs, later models are implemented which have more up to date information on wind and pressure and these have a smaller margin of error. Eventually a decision will have to be made as to whether the likely magnitude of the flooding is enough to warrant the expense and inconvenience of implementing the most severe flood prevention measures. It is always better to be safe than sorry! Figure 6.15 shows an example of how well such models can work. Storm surge modelling is a good, possibly unique example of successful modelling. There will be a chance to do some modelling yourself, albeit much simplified in chapter 8.

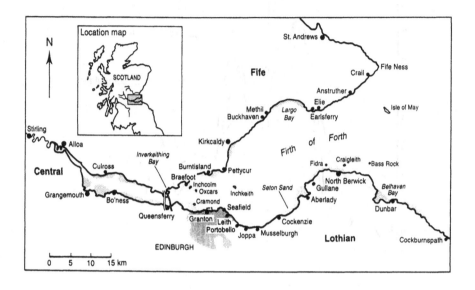

Figure 6.16: The Forth Estuary and the Firth of Forth. Reprinted by permission of the Royal Society of Edinburgh

6.10   Case Studies

6.10.1   The Firth of Forth and Forth Estuary

Estuaries are indeed special places, and deserve special models. There are texts entirely devoted to the physical and dynamical description of estuaries e.g. Dyer (1997). This brief case study in no way competes with such texts, instead we point out some of the key points of modelling this and similar estuaries. The Firth of Forth is a wide estuary located in the south east of Scotland. It is at the mouth of the river Forth, a major Scottish river, but from the road and rail bridges at Queensferry to the North Sea the salinity is virtually the same as the North Sea (see Dyke (1987)). Oceanographically it more closely resembles an inlet than an estuary. In fact the name "Forth Estuary" is usually preserved for that part of the estuary upstream of the bridges to Alloa (see the location map, Figure 6.16 and Webb and Metcalfe (1987)). Because of this geographical split, models of the estuary tend to be built in two parts, one covering the estuary to the east of the bridges and the second to the west of them. There are no complete models of either. Engineering companies have built very localised models for specific purposes, for example managing the environment in terms the siting of marinas and sewage outfalls. What models there are tend not to be available in the public domain.

Models of the estuary upstream of the bridges follows the norm and we have a freshwater outflow over the deeper saline water which is pulled up as a saline wedge. Downstream of the bridges, the freshwater river outflow tend to hug the southern shore, perhaps as a result of the Coriolis effect (see Dyer (1997)). The two layer simple model of Dyke (1980) confirms that it could be the Coriolis effect. Two layer models are however only appropriate in the outer Forth some of the time. There is a seasonal thermocline present only in the late spring and summer. At other times the outer Firth of Forth is well mixed. The equations valid in a rectangular two layered inlet are

$$\frac{\partial u'}{\partial t} - fv' = -g\frac{\partial \eta'}{\partial x}$$

$$\frac{\partial v'}{\partial t} + fu' = -g\frac{\partial \eta'}{\partial y}$$

$$\frac{\partial}{\partial t}(\eta' - \eta'') + h'\left(\frac{\partial u'}{\partial x} + \frac{\partial v'}{\partial y}\right) = 0$$

for the upper layer and a slightly more complicated set

$$\frac{\partial u''}{\partial t} - fv'' = -g\frac{\partial}{\partial x}\left(\frac{\rho'}{\rho''}\eta' + \frac{\rho'' - \rho'}{\rho''}\eta''\right)$$

$$\frac{\partial v''}{\partial t} + fu'' = -g\frac{\partial}{\partial y}\left(\frac{\rho'}{\rho''}\eta' + \frac{\rho'' - \rho'}{\rho''}\eta''\right)$$

$$\frac{\partial \eta''}{\partial t} + h''\left(\frac{\partial u''}{\partial x} + \frac{\partial v''}{\partial y}\right) = 0$$

for the lower layer. The notation is straightforward: a single dash for the upper layer, a double one for the lower layer. (u, v) is the velocity, h the undisturbed layer depth, η the displacement of surface or interface from equilibrium and ρ the density. The deduction from them follows the more general arguments of Gill and Clarke (1974). Namely that the flow can be put in terms of barotropic and baroclinic modes $(u^{(1)}, v^{(1)})$ and $(u^{(2)}, v^{(2)})$. These are related to the variables above via

$$(u^{(1)}, v^{(1)}) = (u', v') + \frac{h''}{h'}(u'', v'')$$

$$(u^{(2)}, v^{(2)}) = (u', v') - (u'', v'')$$

$$h^{(1)} = \frac{h'h''}{h' + h''}.$$

The second baroclinic mode is associated with the baroclinic radius of deformation c_1/f where c_1 is the internal wave speed $\sqrt{(\rho'' - \rho')gh^{(1)}/\rho''}$ and f is the Coriolis parameter. As $\rho'' - \rho'$ is a difference in density and the ratio $(\rho'' - \rho')/\rho''$ about $1/1000$, c_1 is typically one thirtieth of the barotropic value. Hence the Coriolis effect through baroclinicity can be felt on estuarial length scales ($\sim 10km$). The boundary conditions appropriate to a rectangular estuary are that there is no flow through the closed end and that there is no flow through the sides. Theoretical models are very simple and serve only to establish general principals (but see for example Smith (1980) for a not so simple theoretical

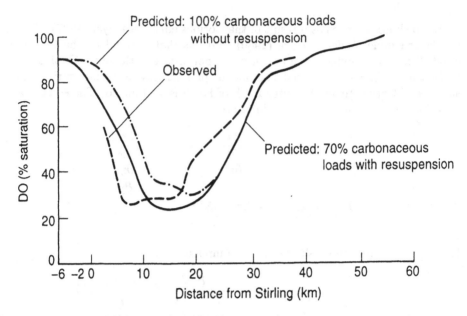

Figure 6.17: Observed and predicted dissolved oxygen (DO) distributions. Tidal range 5.2m, river flow exceedence 80%. Reproduced by permission of the Royal Society of Edinburgh and Dr D.S.McLusky.

model). In order to build a workable numerical model, we would need one that copes with modes. One option is a time splitting technique with a marching method working with both barotropic and first baroclinic modes simultaneously but this will only work if there is no interaction. As more modes are taken into account, this interaction will unfortunately increase and the time splitting technique will not work. Another option is to use a continuously stratified model, neither of these have been done in eastern part of the Forth.

In the west where there is a more conventional estuary there have been many successful modelling studies. In particular one dimensional models which predict dissolved oxygen level and the transport of sediment. These models are data driven with a simple (Fickian) diffusion to simulate spreading. Such models were discussed in Chapter 5 and no further details are given. The output of one such model is given in Figure 6.17. The imperfections of the model can be ascribed to the model assumptions that there is a single source of pollutant and a constant cross-sectional area. Although the second of these could be relaxed, this meant the loss of exact solutions against which to validate the model.

Chapter 7

Environmental Impact Modelling

7.1   The Greenhouse Effect

The greenhouse effect is a well known phenomenon and can be simply explained as follows. As the radiation from the sun makes contact with the earth, some of it is absorbed and some of it is reflected back into space. Surfaces such as the upper surfaces of clouds and snow are particularly efficient at reflection whereas other surfaces such as the sea and forest are not. As the earth is warmed by this radiation, it is emitted back into the atmosphere but at a longer wavelength. Some of this long wavelength radiation is absorbed and then re-emitted by trace gases. It is this transparency of trace gases to (short-wave) incoming radiation contrasting with the blocking of (long-wave) outgoing radiation by the same gases that leads to the warming, commonly called the greenhouse effect. Research indicates that without the greenhouse effect, the temperature of the earth would be 33°C less than the present. The problem now of course that mankind is pumping carbon dioxide and other gases into the atmosphere whilst eliminating large areas of equatorial forest the trees in which provide natural absorbers for these gases. Other effects such as acid rain are also important and are a consequence gaseous discharges from man's industrial activities. Thus the greenhouse effect is enhanced and the earth is heating up. The term "global warming" is the phrase which has been popularised in the press. Most think that global warming will take place and result in an increase in the earth's average temperature, the rate of 0.3°C per decade has been quoted. There is still controversy however and not everyone is convinced of global warming. (See the May 2000 edition of the UK Meteorological Office coffee table journal *Weather* which has letters of interest.) The consequences of global warming are by no means obvious at local scales. For the UK (where this book is being written!) the consensus is for wetter winters, enhanced winds and a sea level rise by about 0.18m by the year 2030. Consequences of this for agriculture, coastal protection water management, river management and ecosystems are uncertain and exercising governments and their various environmental agencies as well as agricultural and other industries. Coastal models of the type outlined in Chapters 2 and 6 can be adapted to modelling some of the consequences of global warming reasonably easily and quickly. For example, an enhanced sea level, wind and rainfall in modelling terms is merely a question of changing some parameters in the model. Other consequences are more difficult to model, for example the erosion of the coastline and changes in animal species. So far in this book no attention has been given to the modelling of the interactions of biological species with others and the environment. Let us rectify this by first looking at how to model ecosystems.

7.2   Ecosystem Modelling

7.2.1   Introduction

An ecosystem is an interconnected biological system involving animals, plants, nutrients and waste products. There are several features to ecosystems that make their modelling very distinctive. The essential feature of modelling as described so far in this text is the writing down of balances. These balances arise form physical laws such as the conservation of momentum (Newton's second law), the conservation of mass, and so on. For modelling ecosystems, there are no such laws. Instead, the understanding of how a particular component of an ecosystem behaves arises from understanding those effects that cause it to grow and those effects that cause it to die. Other effects, such as their direct interaction with another component, or the inclusion of a source of the component itself, can also be incorporated. Each component thus gives rise to an equation. On the left-hand side is the growth rate of the particular component, and on the right are the terms which influence its growth or its decay. In general therefore, we have an equation which has the following structure:

$$
\begin{aligned}
\text{growth rate of component } x \quad = \quad & (\text{a positive constant}) \times x \\
+ \quad & (\text{a negative constant}) \times x \\
+ \quad & \text{source terms} + \text{interaction terms.}
\end{aligned}
$$

The first term on the right-hand side represents growth factors (e.g. feeding, and growth due to internal metabolism of nutrients). The second term on the right-hand side represents the decay rate (e.g. defecation, dissipation by internal metabolism and ultimately death). The third term represents new sources of x (these might be conversions from other components or the mobility of x causing it to migrate into the domain of the problem). Finally, the fourth term represents the fact that what happens to other components can influence what happens to x. This single-variable model, although limited in its applicability, can model certain aspects of the growth of an organism quite successfully. Figure 7.1 shows what is called *logistic growth*. Logistic growth is characterised by an initial exponential behaviour being limited by a ceiling or capacity. As such, although

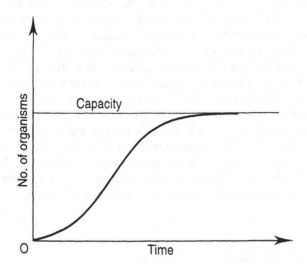

Figure 7.1: The logistic curve

this limiting is deemed to be caused by the organism itself - for example, the self-shadowing of an algal bloom in the upper layers of the sea - it could be caused for example by an (unmodelled) predator species. The fact is, the logistic model is a successful model of organism behaviour even though it is widely recognised that no organism exists in isolation, as such a single-variable model may at first imply. The next step from the logistic growth model is to include, explicitly, the predator species. For example, if x is the biomass or number of individuals of an animal, then obviously if its staple diet, y say, undergoes a drastic increase then x itself will increase. One may thus expect y to appear in an equation governing the growth rate of x, and indeed vice versa. We will meet this again when discussing the simplest ecosystem model, the predator-prey cycle. For now, we have established that each component has an equation, and we have also seen the general form of that equation. One of the most severe criticisms of these models is the plethora of parameters that can be prescribed seemingly at will. It is the values of these parameters that would be changed in order to include the effects of global warming. For example the increase in temperature might increase the metabolism of a species or change the volume of nutrient available. On a wider scale, the increase in sea level will change the salt water - freshwater balance in low lying coastal areas which means entirely restructuring the local ecosystem model.

What we have thus established is an equation for each component and within each equation a right-hand side that contains many parameters that govern the components growth and decay rate, its sources (if any) and its interaction with its fellow components. It is quickly apparent that even the simplest ecosystem can have many equations. This is another distinctive feature of models of ecosystems: lots of equations to accompany the many free parameters already mentioned. In most ecosystems models the equations themselves are assumed linear and are therefore relatively easy to solve, certainly compared with equations that govern fluid flow. This is not because it is certain that the equations are linear, far from it. There is no reason to be more complex than necessary, so linearity will do if it seems to work. It is only in the last twenty or so years that there has been recognition that even a single non-linear term, perhaps xy in the above notation, can lead to the kind of behaviour which has been christened chaos. Chaotic behaviour is easy enough to model and to describe in a general sense; there are many excellent books usually with nonlinear dynamics in the title. As far as we are concerned, it does not seem to model the short term behaviour of biological populations so such models are generally avoided.

7.2.2 Predator-prey ecosystems

The simplest type of model is one that involves a single variable. This variable is usually "population", and the name for this kind of model is a population model. A population of a species can evolve in three ways, provided it is not entirely unpredictable. It can grow without limit, settle down to a steady value (perhaps a zero value) or it can exhibit periodic behaviour. The logistic behaviour mentioned in subsection 7.2.1 is an example of a population that settles down to a value (its capacity). The right-hand side of the single equation that governs the change of population can be so structured that any one of these outcomes are possible. Mention ought to be made here of stochastic modelling, since population studies are its most straightforward manifestation. The word stochastic implies that there is a statistical or probabilistic element to the modelling. In population studies, several algorithms based on random walk techniques of the type already met in Chapter 5 are possible. In these, there is a random element to the precise value of the population at any instant, and although averaging procedures can regain the previously mentioned underlying trends, only estimates of actual population are possible. In more sophisticated stochastic population models, the nature of the randomness itself can influence long-term growth or decay. More will be said about statistical modelling in the next chapter. The simplest biological population model involving two species is called a predator-prey model. The equation for each species takes a form that permits a closed graphical solution, as shown in Figure 7.2. It is instructive to look at this graphical solution alongside the equations themselves, which are written in the form:

$$\text{Rate of change of } x = \frac{dx}{dt} \;\; = \;\; ax - cxy,$$

$$\text{Rate of change of } y = \frac{dy}{dt} \;\; = \;\; -by + dxy.$$

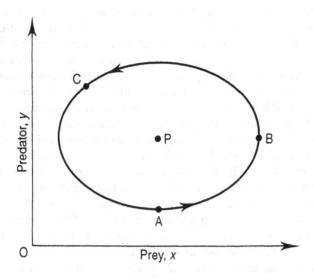

Figure 7.2: The predator-prey model, no damping.

The mathematical solution of these two equations is possible in closed form and is the origin $(0,0)$ which *is* a solution albeit a trivial one, plus curves of the form

$$a \ln y + b \ln x - cy - dx = \text{constant}.$$

It is a representative of one of these curves that is drawn in Figure 7.2. It will be noticed that the values $x = b/d$, and $y = a/c$ correspond to zero growth of both predator and prey and the point

$$\left(\frac{b}{d}, \frac{a}{c}\right)$$

is at the "centre" of all closed curves of the type displayed in Figure 7.2. It is useful to describe qualitatively what happens to both species as a typical solution curve is traversed. The prey (strictly, either the biomass of the prey or the number of individuals), labelled x, naturally grows (probably by eating grass, which is not included in the equations). The rate of this growth is represented by the constant a. On the other hand, the predator, labelled y, eats x at a rate governed by the magnitude of the constant c. The total rate of change of x is governed by these two competing effects. The predator naturally dies at a rate governed by the constant b (the death rate). On the other hand, it grows from eating x at a rate governed by the constant d. Note that although c and d represent the same process (y eating x) they are different because c denotes the effect on the prey, whereas d denotes the effect on the predator. For this simple model, arguments can be proposed that can justify putting $c = d$.

The cycling depicted in Figure 7.2 can be explained in words as follows: as the number of predators is low, to the left of point A on the curve, the prey can increase by grazing without fear of being pounced upon. As the number of prey reaches a maximum (point B), the population of predators also thrives due to the plentiful food supply. The inevitable consequence of this is a decrease in prey until (point C) they become scarce enough to diminish the predator population. Once the predator population has reached a low enough value, the prey thrives and the whole cycle begins again. This is the simplest model. If extra terms are added to the right-hand sides of these equations so that more sophisticated eating habits and more complex relationships between the number of predators, the number of prey and growth rates represented, then it is possible for the curve to spiral inward towards the point $(\frac{b}{d}, \frac{a}{c})$. Such a point is called an equilibrium point as both rates of change are zero there. It represents a stable point at which fixed numbers of predators and prey can live in perpetual harmony. When more variables are involved, a great variety of different stable, unstable and oscillatory states might well be possible. For a catalogue of possible states in two dimensions, see books on nonlinear dynamics, the book by Berry (1996) is a user friendly simple introduction.

Let us now examine one practical ecosystem model which helps to explain some of the well known characteristics of the biochemistry of the surface waters in the North Atlantic that were investigated during BOFS (the Biogeochemical Ocean Flux Study 1988-92, a UK national research programme funded by the Natural Environment Research Council and forming the UK end of JGOFS, the Joint Global Ocean Flux Study).

7.2.3   Modelling a real marine ecosystem

There are many biological and chemical signals in the ocean which one can attempt to model. The decision has been made here to look at a model which examines the role of phytoplankton in the carbon balance in the North Atlantic Ocean, particularly at the surface. This model is selected because of its relevance to global warming, one of the main concerns of today. Models which are published in the international scientific literature are very well researched, and usually stem from a history of similar models together with knowledge gained from many observations and experiments. What follows is therefore not a justification of all the modelling details in terms of their biology and chemistry. Instead, a documentary style account of the various balances is given, together with the model equations, graphical results and a visual comparison with observations.

The model of Taylor *et al.* (1991) is essentially a point model. That is, only a single point in space is considered. There is no horizontal advection or transportation of material. The value of such models lies in their ability to isolate biochemical effects from hydrodynamic effects. Point models of biology and chemistry can be embedded in hydrodynamic models at a later stage, since only then can parameters appropriate to the interaction between fluid flow and biochemical effects be added. At the time of writing, very little work has been done on the interface between hydrodynamics and biochemistry in the sea. The model under the microscope here certainly contains no hydrodynamics. What

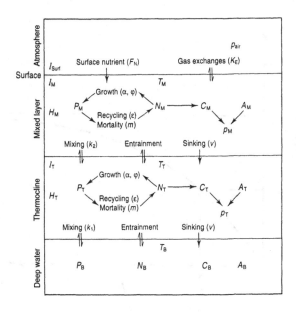

Figure 7.3: A marine ecosystem model. A, alkalinity; C, total inorganic carbon concentration; P, phytoplankton abundance; N, nutrient concentration. From Taylor *et al.* (1991). Reproduced with permission from American Geophysical Union, Washington, USA.

it does consider, however, is some depth dependence. Although no horizontal structure is present, there is a mixed layer which interacts with the atmosphere above and a thermocline layer below. In turn, this thermocline layer also interacts with the deep water below it. The main biology and chemistry takes place in the mixed layer and thermocline regions. In each layer, there are four variables making eight in all. These variables are phytoplankton concentration, nutrient concentration, concentration of dissolved carbon and alkalinity. At its heart, the model therefore has eight rate equations which relate rates of change of these quantities to processes such as diffusion between layers, sinking rate of phytoplankton, carbon recycling efficiency, etc. None of the right-hand sides of these rate equations contains any non-linear terms, and hence there is no possibility that chaos will manifest itself. Models such as this are difficult to grasp from a cold start. Diagrams that illustrate the processes, such as that given in Figure 7.3, help us to understand what is taking place. For those who understand mathematics, this is a poor substitute for the precision of equations which come later, but it is a good place to start. The most important input to a model such as this is nutrient, and it is thus crucial that this variable behaves realistically. Most researchers in this field are satisfied that nutrient obeys the Michaelis-Menten relationship. There is no rigorous justification for this in terms of well defined laws, although the following provides a justification of sorts. If N denotes nutrient concentration and α denotes a value of

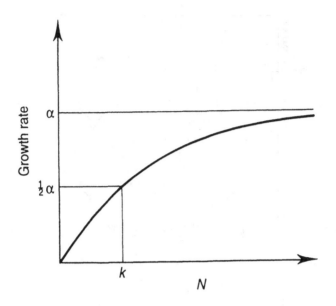

Figure 7.4: The Michaelis-Menten growth relationship.

growth deemed to be an ideal maximum value reached when conditions pro-
mote maximum nutrient uptake, then the growth of nutrient is governed by the
expression:

$$\frac{\alpha N}{k + N}.$$

This expression has the following essential properties. When there is no nutrient,
there is zero growth. For very large N, the growth approaches the maximum
(ideal) value α. For values in between, the growth is between zero and α, but
it is always bigger for bigger N. Finally, the constant k is available for fitting
to the data. In fact, if $N = k$, then the growth rate is $\frac{1}{2}\alpha$ (some readers may
spot a loose analogy with radioactive half life here). Figure 7.4 gives a general
graph of the Michaelis-Menten relationship.

The model of Taylor *et al* (1991) includes eight variables, phytoplankton P,
nutrient N, concentration of total dissolved carbon C and alkalinity A in each
of two layers. The variables in the mixed layer have a suffix M and those in the
thermocline layer a suffix T. The model itself is assumed to be governed by the
following set of eight equations:

$$\frac{dP_M}{dt} = \left[\alpha(I_M, T_M)\phi(N_M) - m - \frac{\nu}{H_M}\right] + k_2\frac{P_T - P_M}{H_M}$$

$$\frac{dP_T}{dt} = [\alpha(I_T, T_T)\phi(N_T) - m]\,P_T + \nu\frac{P_T - P_M}{H_T} + k_1\frac{P_B - P_T}{H_T} - k_2\frac{P_T - P_M}{H_T}$$

$$\frac{dN_M}{dt} = -\gamma\,[\alpha(I_M, T_M)\phi(N_M) - \epsilon m]\,P_M + k_2\frac{N_T - N_M}{H_M} + (1 - \theta)\frac{F_N}{H_M}$$

$$\frac{dN_T}{dt} = -\gamma[\alpha(I_T,T_T)\phi(N_M) - \epsilon m]P_T + k_1\frac{\theta N_B - N_T}{H_T} - k_2\frac{N_T - N_M}{H_T}$$

$$\frac{dC_M}{dt} = -[\Gamma_C\alpha(I_M,T_M)\phi(N_M) - \Gamma'_C\epsilon_C m]P_M + k_2\frac{C_T - C_M}{H_M} - k_H\frac{\rho_{air} - \rho_M}{H_M}$$

$$\frac{dC_T}{dt} = -[\Gamma_C\alpha(I_T,T_T)\phi(N_T) - \Gamma'_C\epsilon_C m]P_T + k_1\frac{C_B - C_T}{H_T} - k_2\frac{C_T - C_M}{H_T}$$

$$\frac{dA_M}{dt} = \Gamma_A\gamma[\alpha(I_M,T_M)\phi(N_M) - \epsilon m]P_M + k_2\frac{A_T - A_M}{H_M}$$

$$\frac{dA_T}{dt} = \Gamma_A\gamma[\alpha(I_T,T_T)\phi(N_T) - \epsilon m]P_T + k_1\frac{A_B - A_T}{H_T} - k_2\frac{A_T - A_M}{H_T}.$$

These equations are very full of parameters which is typical of ecosystem models. There are only the four variables, but the depth of each layer H enters the equations, the suffices M and T refer to the mixed and thermocline layers respectively. The k parameters indicate mixing parameters and the suffix B denotes the bottom layer beneath the thermocline. In this layer the variables are taken to have fixed prescribed values. The mortality m is found explicitly in the phytoplankton equations and prefixed by various fractions (indicated by ϵ with or without suffices) elsewhere. It will also be noticed that the growth of each variable is dependent on a parameter α which changes with irradiance I and temperature T but is often taken as constant. This is because knowledge of how α actually varies is particularly scant, not because we are sure it is constant. The other parameters Γ (with various suffices) are reduction factors and in one instance (Γ_A) merely a conversion factor from millimoles nitrogen per cubic meter to moles kilograms! ρ is the density and ϕ a function governing how the growth of all variables depends on nutrient. Let us discuss this model. The discussion does not in any way presume that you are up to speed with these equations; it can be read separately.

In this model, nutrient limitation is represented by the Michaelis-Menten relationship. The logic of this is seen once it is realised that phytoplankton grow by photosynthesis, and photosynthesis is directly proportional to irradiance. Therefore the presence of nutrient in the ecosystem will be limited by the amount of this sustaining irradiance. However, this global limiting via the Michaelis-Menten relationship will only apply to the total nutrient and at the microscopic level a more complicated relationship governing the nutrient growth in each layer is appropriate. In this model, once the amount of nutrient reaches a certain level, the Michaelis-Menten limiting relationship dominates other effects. The rate of change of phytoplankton in the mixed layer is governed by growth through nutrient absorption, a loss of phytoplankton through it sinking into the thermocline layer below, an additional death rate loss plus a diffusion, proportional to the difference between the concentrations of phytoplankton in the mixed and thermocline layers. Of course, this latter diffusion could act either as a source of phytoplankton (if the thermocline population exceeded the mixed layer population) or as a sink (if the reverse were true). In the thermocline layer, the negative of this diffusion process is present because the net effect of diffusion on the phytoplankton is zero, as we saw in Chapter 5; diffusion spreads

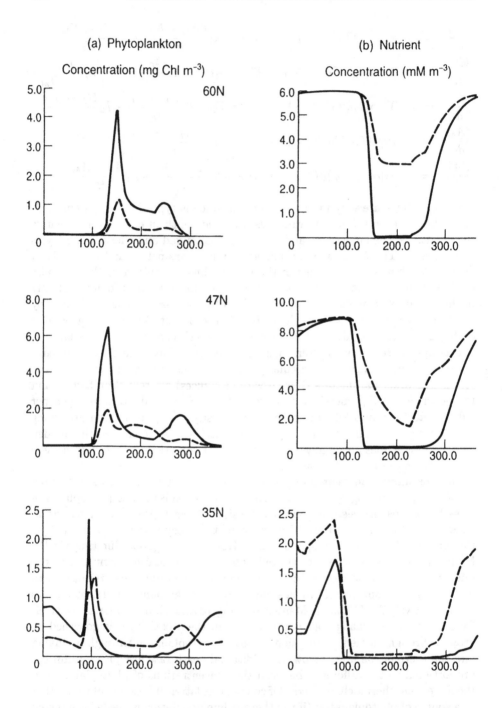

Figure 7.5: Model predictions of Taylor *et al.* (1991). In these runs the nutrient is diffusing up from depth. Solid lines, mixed layer values; dashed lines, values in the thermocline. From Taylor *et al.* (1991). reproduced with permission from American Geophysical Union, Washington, USA.

out populations, but it cannot create or destroy them. In this thermocline layer, the phytoplankton concentration also grows through nutrient absorption and is depleted by mortality and through sinking. There is an additional diffusion via the deep layer which could be a source or a sink (as above) but is usually the latter since phytoplankton are not abundant in the deep sea. Below concentrations where Michaelis-Menten limitation is dominant, the nutrient concentration in each layer is governed by similar competing mechanisms. First, there is a recycling term due to the photosynthetic effect in the phytoplankton. This is subject to an efficiency index - after all the phytoplankton keep some nutrient to live, and only give up the surplus. Whatever nutrient diffuses into the mixed layer also diffuses out of the thermocline layer and vice versa. Diffusive interaction between the thermocline layer and the deep ocean also takes place. For the nutrient concentration, there are also inputs from the deep and also from the atmosphere. These external inputs are crucial to this model since one of its prime functions is to predict seasonal changes; hence seasonal nutrient input is a prime driving mechanism. One of the main outputs of the model is total dissolved carbon. The model, besides tracking this dissolved carbon through its interaction with the growing, dying and sinking phytoplankton also includes a term which represents the transfer of carbon dioxide into (and perhaps out of) the sea surface. There are also the standard diffusion terms which take the same forms as previously described. The final variable is alkalinity, the value of which merely depends on the phytoplankton processes plus diffusion in each layer. In summary, therefore, the main dynamics of the model are phytoplankton growth, sinking and death, the interaction of this with nutrients, dissolved carbon and alkalinity, including all types of diffusive effects, and seasonal inputs of nutrient.

The key to whether a model such as this is a good one lies in how it compares with observation. Taylor *et al.* (1991) had the good fortune to be sitting on top of very good sets of data which had just emerged from cruises (1989) arising out of the Biogeochemical Ocean Flux Study. These data enabled the parameters of the model to be assigned with perhaps a little more than the normal low level of confidence. Mention here ought to be made of the consequences arising from the absence of horizontal variation in the model. The model operates quite well at some latitudes, and fails miserably at others. The authors applied the model at several latitudes corresponding to where there was sufficient supply of data (models such as these are very data-hungry, unlike the hydrodynamic models of Chapters 2 and 6). They concluded that it was only appropriate to examine how the model performed at 60°N, at which latitude the phytoplankton reside primarily in the surface mixed layer throughout the year, and the main seasonal variations are in agreement with observations. Figure 7.5 shows the model predictions. Fortunately, weather ship *India* was located at 59°N, 19°W for a number of years gathering data that can be used to help validate models such as this. The BOFS cruises collected measurements of chlorophyll-*a* (a measure of phytoplankton concentration) and the partial pressure of carbon dioxide at 60°N, 20°W over a five-day period. The chlorophyll-*a* measurements could be compared directly with predictions (Figure 7.6). The measurements of partial pressure of CO_2 could also be compared with the model output once

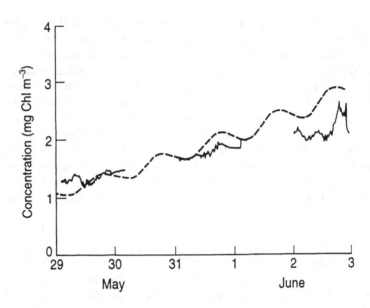

Figure 7.6: Chlorophyll-*a* concentrations. Solid lines, measured values; dashed lines, model predictions. From Taylor *et al.* (1991). Reproduced with permission from American Geophysical Union, Washington,USA.

the total dissolved carbon and alkalinity were related to this partial pressure via the standard chemical equations of sea water. The comparison is shown in Figure 7.7. Another success of the model was the insight that was gained into coccolithophore blooms. Coccolithophores are a group of phytoplankton that have hard calciferous ($CaCO_3$) shells, are widespread in the oceans and provide a significant source of deep-sea carbon. Blooms of a species of coccolithophore called *Emiliana huxleyi* occur in mid-latitudes (45°-65°) in spring and summer, so much so that the ocean can appear milky white due to light back-scattering. Taylor *et al.* (1991) used their model successfully to simulate various aspects of this coccolithophore bloom.

Let us now discuss the sensitivity and limitations of this model. The authors of this model, in recognition of the absence of lateral transfer in the model, added it crudely in one experiment via a simple northwards flow. Apart from an earlier spring bloom, there were no changes of major note. The sensitivity to changes in carbon dioxide exchange across the air-sea interface was also examined. Some changes, particularly in summer values, were noted. Finally, the model was quite sensitive to the value of phytoplankton mortality and other parameter values which link phytoplankton abundance to nutrient.

It is quite typical of models such as this one to be quite sensitive to key parameter values. As the number of observations increases, so the estimates of the values of these key parameters can be improved. It must be remembered, however that many parameters (for example, the diffusion coefficients) are at

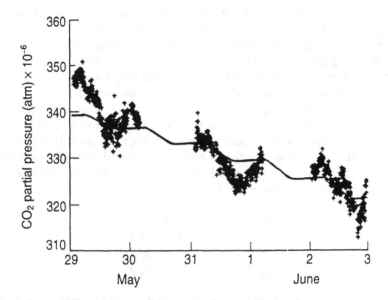

Figure 7.7: Partial pressure of CO_2. Crosses, measured values; full line, model predictions. From Taylor *et al.* (1991). Reproduced with permission from American Geophysical Union, Washington, USA.

best of doubtful pedigree and will always have a high degree of uncertainty associated with them. If a particular model prediction is very sensitive to such parameters, then the model output will be unreliable. When reading the details of an ecosystem model, always look for sensitivity analysis and arguments to back up the modelling predictions. These arguments should be based on a sound knowledge of the biological and chemical processes that are being modelled. A model which is given without its context is virtually worthless.

7.2.4 Other ecological models

The literature on ecosystems modelling is quite new in terms of modelling marine processes, but it is already extensive. There are many different kinds of approaches, approaches so different that it is often difficult for the lay reader to see any connection between them. This reflects the lack of an equivalent to the 'Newton's second law of motion' starting point to models of physical systems. One interesting new approach is to use simple equations, usually modified Lotka-Volterra, and plot solutions graphically. This approach, however, is limited to models that contain only at most three independent variables, so we will not dwell on details here except to say that there is a strong possibility of relating such models to modern developments in dynamic system theory, which places no restriction on the number of independent variables (apart from the inability to clearly visualise more than three).

Instead, let us look at an attempt to model the deep chlorophyll maximum (DCM). Such a feature has been known about for at least 50 years, but it is only recently that its importance to the biological productivity of the region in which it occurs has been explicitly stated. The model detailed here is taken from the work of Varela *et al.* (1994). Before detailing the biological model, we need to specify the physical model in which it is embedded. This time let us start with hydrodynamic equations of the type found in Chapter 2. The model here assumes no pressure gradients as these can be superimposed as large scale features. The authors also propose a more complex than usual frictional model that takes the form of a turbulence closure scheme; details can be found in Chapter 4. There is no need to include the troublesome non-linear terms, so the governing equations are of the form:

$$\frac{\partial u}{\partial t} - fv = \frac{\partial}{\partial z}\left(\overline{v}\frac{\partial u}{\partial z}\right),$$
$$\frac{\partial v}{\partial t} + fu = \frac{\partial}{\partial z}\left(\overline{v}\frac{\partial v}{\partial z}\right),$$

where, as usual, u and v are the easterly and northerly components of the velocity, f is the Coriolis parameter, z is vertically upward, t is time and $\overline{v}$ is an eddy viscosity. It is this latter eddy viscosity that is treated rather differently from earlier models in that it is related to thermodynamic variables (instead of being assumed constant, as is usually the case). First, there are diffusion equations for salt and heat which introduce the standard diffusion coefficients (for simplicity these diffusion coefficients are assumed equal). The eddy viscosity is related to the turbulent kinetic energy (k) and the mixing length (l_m) via the simple relationship:

$$\overline{v} = \frac{1}{2}\sqrt{kl_m}$$

which is due to Mellor and Yamada (1974). The turbulent kinetic energy is then related to the buoyancy produced by the presence of heat and salt through a number of quite straightforward equations which include the quantities the Richardson numbers and the Brunt-Väisälä frequency. These terms are, hopefully, familiar from Chapter 2. The difference between the standard Richardson number and the flux Richardson number is that to obtain the latter, multiply the former by the thermal diffusivity divided by the eddy viscosity. The relationships:

$$R_i = \frac{N^2}{(\partial u/\partial z)^2 + (\partial v/\partial z)^2},$$

where R_i is the (standard) Richardson number, and N is the Brunt-Väisälä frequency, given by:

$$N^2 = -\frac{g}{\rho}\frac{\partial \rho}{\partial z},$$

relate these dimensionless numbers to the velocity and the density. The density is in turn related to the salinity and temperature through an equation of state.

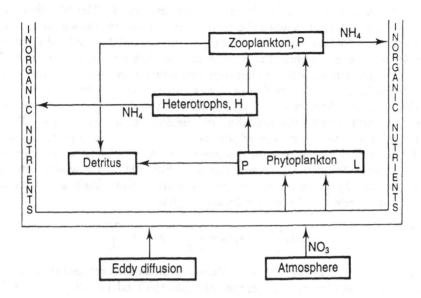

Figure 7.8: The trophic flow chart. From Varela *et al.* (1994). Reproduced by permission from the author.

If some of the mathematical details above passed you by, the essential point to grasp is that the following biological modelling is soundly based on an up-to-date model of the underlying circulation. The biological variables used in this model are ammonia NH_4, nitrate NO_3, two kinds of phytoplankton (L for large and P for small), and heterotrophs, H (heterotrophs are creatures that get all their carbon from consuming green matter, in this case phytoplankton). The equations that they obey are of the usual variety, that is, they are rate equations with the rate of change of any particular variable being equal to (read 'governed by') uptake of the other variables, excretion and grazing. For completeness, and because we have avoided giving examples of the previous simpler model, we give these biological equations below. In what follows no understanding of any of the mathematical material will be assumed.

$$\frac{\partial(NO_3)}{\partial t} = \frac{\partial}{\partial z}\left(\lambda\frac{\partial(NO_3)}{\partial z}\right) - U(NO_3),$$

$$\frac{\partial(NH_4)}{\partial t} = \frac{\partial}{\partial z}\left(\lambda\frac{\partial(NH_4)}{\partial z}\right) - U(NH_4) + \psi_Z + \psi_H,$$

$$\frac{\partial L}{\partial t} = \frac{\partial}{\partial z}\left(\lambda\frac{\partial L}{\partial z}\right) - s_1 + U(L) - G_{ZL},$$

$$\frac{\partial P}{\partial t} = \frac{\partial}{\partial z}\left(\lambda\frac{\partial P}{\partial z}\right) - s_2 U(NO_3 P) - G_{HP},$$

$$\frac{\partial H}{\partial t} = \frac{\partial}{\partial z}\left(\lambda\frac{\partial H}{\partial z}\right) - s_3 + G_{HP} - G_{ZH} - \psi_H.$$

All these equations are, as we have said, rate equations. The left-hand sides of these equations give simply the rates of growth of nitrate, ammonia, large and small phytoplankton and heterotrophs, and the right-hand sides express how these rates are governed by various agents. The uptake is denoted by U, where the parentheses following indicate that which is uptaken. The composites NO_3P and NH_4P refer to uptakes of nitrate and ammonia, respectively by small phytoplankton. The terms s_1, s_2 and s_3 refer to the three sinking rates of large phytoplankton, small phytoplankton and heterotrophs, respectively. The G terms refer to grazing by zooplankton, heterotrophs and phytoplankton (Z, H and P respectively), and ψ indicates ammonia excretion. The various uptakes are calculated through a knowledge of available light and available nitrogen using empirically derived relationships, which the authors freely admit are not universally accepted. The sinking terms are given by:

$$s_1 = v_1 \frac{\partial L}{\partial z}, \quad s_2 = v_2 \frac{\partial P}{\partial z}, \quad s_3 = v_3 \frac{\partial H}{\partial z}.$$

Relationships such as the Michaelis-Menten formula that represents light limitation by self-shadowing are incorporated into the model through the uptake, sinking and grazing terms. The trophic model in terms of a flow diagram is shown in Figure 7.8. Note that the presence of zooplankton is not modelled explicitly. The results of this particular model are quite good in that the predicted concentrations of nitrate seem to follow observations (see figure 7.9). Of prime interest, however, is how the model predicts the deep chlorophyll maximum (DCM) which, after all, was the driving force behind the model. In order to see how chlorophyll enters the model, we note that light attenuation, I_z, at a depth z is governed by the expression:

$$I_z = I_O \exp(-(k_w + k_c + k_d)z),$$

where k_w is pure water extinction, k_d is the extinction due to factors such as organic matter and locally determined factors, but k_c is the all important extinction hue to chlorophyll-a concentration for which a linear function is assumed for simplicity. I_O is the value for surface incident light. This model then is capable of predicting the distribution of chlorophyll-a with depth; Figure 7.10 shows one such profile using parameters appropriate to the southwest Sargasso Sea.

It can be seen that this model is successful in predicting the observed chlorophyll maximum. However, one always needs to remind oneself that models such as this are based to some extent on empiricism and have a number of so-called free parameters that can be tuned so that the output fits certain observations. Good fit to data should therefore only be one factor that is examined. An equally important question to ask is whether the model correctly mimics other aspects of the physics, chemistry and biology of the sea. We have not given enough details of this particular model for the reader to answer this question, but in order for a paper to be published in the international literature, peer review should ensure that the model contributes a new insight into the prediction

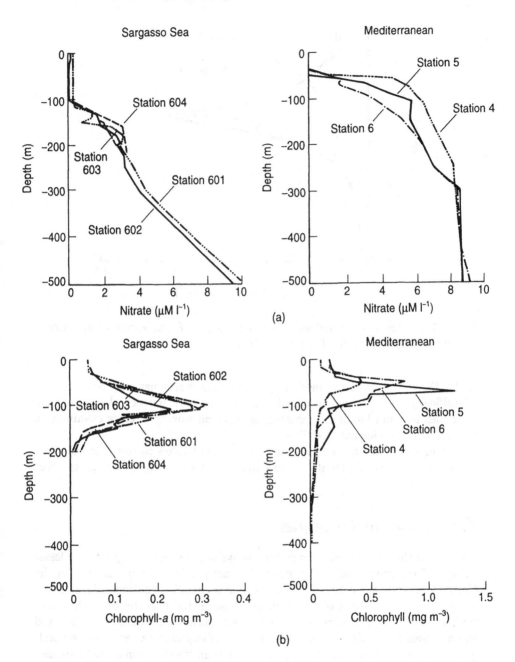

Figure 7.9: Model predictions: (a) nitrates; (b) chlorophyll-*a*. From Varela *et al.* (1994). Reproduced by permission from the author.

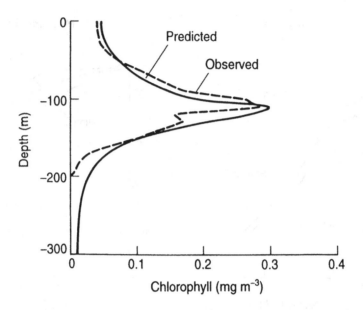

Figure 7.10: The chlorophyll variation with depth. From Varela *et al.* (1994). Reproduced by permission from the author.

of the DCM, and that the modelling is accurate mathematically and credible scientifically. This model, combining as it does a turbulence closure scheme with chemical and biological variables is new and worthy of close examination by students of ecological modelling.

Let us now move away from purely ecological models to consider the other major consequence of global warming due to climate change, that is its effect on the coast.

7.3 Coastal Changes

Changes in the shape of coastlines has always taken place to a greater or lesser extent. There are places in the world that are particularly vulnerable. In the United Kingdom, the coast of Norfolk has been changing continuously, so much so that the entire village of Dunwich has disappeared into the North Sea over the last few thousand years. Normally it is a storm that causes a sudden land slip and over the years the cliff edge gets inexorably closer to vulnerable buildings which are vacated and lost or in some circumstances demolished carefully brick by brick and resurrected in a safer place. Such was the fate of an historic lighthouse on the Sussex coast. Of course, in other places the opposite is happening. In South Yorkshire, just north of the River Humber, the coast north of Spurn Head is being built up and once coastal towns and villages now find themselves inland. The processes that cause such changes are not completely understood, but there are models that incorporate the essential features and it is useful to summarise these features before describing models. The wave climate is crucial, especially its extreme manifestation under storm conditions. Local tides and other currents need to be modelled as best as possible. Finally, the help of geologists is required to gauge the composition of the coast. Sand and silt beaches, and cliffs composed of easily erodable rocks such as limestone or soft sedimentary rocks are vulnerable. In East Anglia for example, the annual budget of sand being eroded from the north Norfolk cliffs at Overstrand is about $400m^3$ per year. This material mostly ends up replenishing sandy holiday beaches further south.

As the climate becomes warmer and more energetic, such processes will speed up so it becomes ever more important to understand and model them. Only then might something be done to control or at least minimise the effects of costly erosion.

7.3.1 Modelling Erosion

The main difficulty with modelling erosion looks insurmountable. The important mechanisms that control the erosion process lie precisely in the interface between a violently moving sea and a soft moving land. This is notoriously difficult to model. Indeed, even assuming a quiet sea the discussion of Chapter 4 shows that near the bed very small length scales are required to capture the physics. These small scales are at odds with the kind of larger scale finite difference models adopted in coastal sea modelling. There is thus something of an *impasse*. When the sea is anything but quiet, which is of interest here, there must be little hope indeed of a good model. Nevertheless, some attempts have to be made to model the effects of a violent nonlinear sea with its foaming breaking waves crashing against and undercutting sea cliffs. The secret lies in averaging over time in order to forecast long term trends, remembering that this is the long term goal. Before considering waves, let us start by looking at currents.

7.3.2 Real Currents

One common assumption for currents is that close to the sea bed they have an approximately logarithmic profile. This was alluded to in Chapter 4. The horizontal current u is a function of height above the sea bed z, an indicative roughness length z_0 and a "friction velocity" u_* which is approximately $\sqrt{\tau_0/\rho}$ where τ_0 is the bed shear stress. This profile is given by

$$u(z) = \frac{u_*}{\kappa} \ln\left(\frac{z}{z_0}\right) = \frac{1}{\kappa}\sqrt{\frac{\tau_0}{\rho}} \ln\left(\frac{z}{z_0}\right)$$

where κ is von Karman's constant usually attributed to be 0.4. Estimates of z_0 usually rely on classical experiments done in the 1930s by J Nikuradse. However, z_0 has been related to the viscosity of sea water ν (not eddy viscosity; we are firmly in a laminar boundary layer this close to the bed). The constant k_s

is called the Nikuradse roughness which is directly related to the grain size. Experimental fit gives

$$z_0 = \frac{k_s}{30} \left[1 - \exp\left(-\frac{u * k_s}{27\nu} \right) \right] + \frac{\nu}{9u_*}.$$

The expression $(\frac{u_* k_s}{\nu})$ is a Reynolds number (ratio of inertia to viscous terms) which is traditionally the important dimensionless ratio in a fluid lacking waves or the influence of Coriolis effects. Some simplification of this formula for certain ranges of Reynolds number is possible, and the more specialist text of Soulsby (1997) is recommended for further details.

Power laws are prevalent in the coastal engineering literature. For example, the formula

$$u(z) = \left(\frac{z}{0.32h} \right)^{1/7} \bar{U} \qquad 0 < z < 0.5h$$

has been successfully employed to model a tidal current in the bottom half of a flow where the depth of water is h. In the top half, the constant value $1.07\bar{U}$ suffices. (Here, $\bar{U}$ is the depth mean tidal flow.) The one seventh power law is only justified by fitting a power law through a considerable amount of data taken from a variety of coastal seas and estuaries. Fitting a straight line to a log-log plot inevitably produces a power law (see Chapter 8), and the exponent $1/7$ has a long pedigree from mechanical engineering (Schlichting (1975), p600). The logarithmic law can be derived from dimensional analysis in that the velocity gradient can be deduced to be proportional to a representative velocity (u_*) divided by the vertical co-ordinate z which is a (the only) representative length scale. Integration of this yields the logarithmic law. However it is freely admitted that this is not an area where the science is exact; remind yourself of what we are trying to model here!

Currents are of course important, but so are waves and in order to assess environmental impact, both need to be considered together.

7.3.3 Real Sea Waves

In Chapter 6, the rudiments of waves were introduced. Of course in a real sea there are very few pure sinusoidal waves. The waves that are visible on the surface of the sea certainly exhibit the to and fro motion one thinks of as an essential characteristic, but their form is usually complicated, particularly in stormy conditions. A single record of a sea wave will have an appearance like Figure 7.11. The distribution of wave heights within such a time series is often assumed to obey the following rule: The probability that a particular wave height h_w exceeds a given prescribed wave height H_w is

$$e^{-2(H_w/H_s)^2}$$

where H_s is the significant wave height (the average height of the highest one third of the waves). This expression is the cumulative distribution which is the

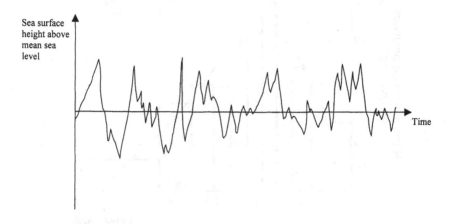

Figure 7.11: A Time Series

integral of the corresponding probability density function

$$p(h) = \frac{2h}{H_{rms}^2} e^{-(h/H_{rms})^2}$$

where H_{rms} is the root mean square wave height of all the waves. This is the Rayleigh distribution. Practically, there will only be a finite number of waves and the graph of p against h (Figure 7.12) is a "line of best fit" through a wave record analysis. The total number of waves in a particular record is N_w say and assuming a Rayleigh distribution it is possible to find estimates of various parameters. For example, the maximum wave in the record is estimated by

$$H_{max} = H_s \left(\frac{1}{2}\ln N_w\right)^{1/2} \sim 1.6 H_s \qquad \text{for a typical 20 minute wave record.}$$

In turn we can also deduce that $H_s = H_{rms}\sqrt{2}$. Unfortunately the Rayleigh distribution only really applies to a limited range of linear (sinusoidal) waves and although it can still be used successfully for swell waves and some wind waves of the variety that might cause long term erosion, it is not good for storm waves. For example, measurements of real waves indicate that $H_s \sim 1.42 H_{rms}$ for swell waves which is in accord with the Rayleigh distribution, but give $H_s \sim 1.48 H_{rms}$ for storm waves. Clearly, more sophisticated spectra are required. Before moving on, it is beneficial to spend a little time understanding this concept of a wave spectrum. The underlying assumption behind wave spectra is that a real sea with all its complexity can be expressed as a sum, perhaps an infinite sum of sinusoidal waves. For those who know about Fourier series, it

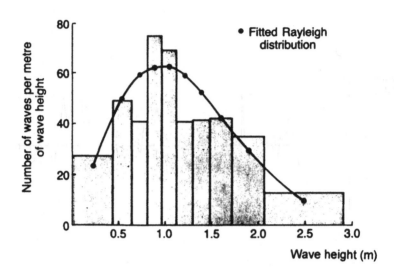

Figure 7.12: The Rayleigh Distribution superimposed on a wave record analysis

will come as little surprise that this is possible. For those who do not, the idea is that *any* periodic signal can be decomposed into a sum of sinusoidal signals of differing frequencies. An extension of this is that any signal whatsoever is composed of a spectrum of frequencies. This might seem preposterous, but one quite convincing analogy is that a complicated television picture can be generated through the careful combination of just three primary colours. For signals however there are no primary frequencies and many are required to generate a given signal. It is tempting to believe that the spectrum of a signal is merely a function which expresses for each frequency the intensity of that sinusoid required successfully to generate the original signal. This however is not quite the case. The reason is because signals do not start or finish but are infinitely long and this gives convergence problems. A single wave if perfectly sinusoidal will have a form

$$\eta = a\cos(\omega t + \epsilon)$$

where a is the amplitude of the wave, ω its frequency and ϵ the phase. A whole train of these waves will take the form

$$\eta = \sum_{n=0}^{\infty} a\cos(\omega n t + \epsilon_n).$$

The infinity is in reality replaced by a very large integer. The right hand side is almost a Fourier series and can be made precisely so with some mathematical tidying up. This is not done here. Fourier series can emulate any time series in theory, but in practice other considerations dominate. The energy of a sinusoidal wave is proportional to half of the square of the amplitude. For the above wave train therefore the energy is E where

$$E = \frac{1}{2}\rho g \sum_{n=0}^{\infty} a_n^2$$

and will in general be infinite. The dimensions of E are in fact MT^{-2} which represents energy *per unit area of sea surface*. Hydraulic and Coastal Engineers will be more familiar with the expression for the single sinusoidal wave of the form $\frac{1}{8}\rho g H^2$ where $H = 2a$. The wave train comprising infinitely many waves will, however it is expressed remain stubbornly infinite in most cases. One of the most successful ways of overcoming this problem is to define a function called the autocovariance which is derived directly from the signal. This function measures agreement between two portions of a signal and decays to zero for large times (gaps between the portions). As it is formed by multiplying one part of the signal with another, it has a similar form to the energy and can be made to have the same dimensions if multiplied by ρg. Indeed it is common parlance in Electronic Engineering to call the time integral of the square the "energy", so if $x(t)$ is a time series of the type depicted in Figure 7.11 then the integral

$$\int_{-\infty}^{\infty} [x(t)]^2 dt$$

is the energy. If $x(t)$ is a length, then this integral will have dimension L^2T, remembering the integration with respect to time. As true energy per unit area has dimension MT^{-2} one needs to multiply this by ρg and divide by time, perhaps the duration of the record, in order to be dimensionally consistent. It is the Fourier decomposition of the autocovariance of the signal that forms what is called the spectrum. This is done by using Fourier transforms, and we briefly show this. If a signal (time series) $x(t)$ is a continuous function of time, then perform the integral

$$X(i\omega) = \int_{-\infty}^{\infty} x(t)e^{-i\omega t} dt.$$

This integral encapsulates the behaviour of the signal as a function of frequency, ω. The presence of $i = \sqrt{-1}$ should not trouble you, but the possible convergence problems associated with the infinite integrals are the subject of modifications that we return to later. The function X is the Fourier transform of x. It can be shown by double application of the Fourier transform that

$$\int_{-\infty}^{\infty} [x(t)]^2 dt = \frac{1}{2\pi} \int_{-\infty}^{\infty} |X(i\omega)|^2 d\omega.$$

The quantity $|X(i\omega)|^2$ is called the *energy spectral density* of the signal and the above relationship is Parseval's theorem. The appearance of 2π is unavoidable and is due to the period of sine and cosine functions. In fact this factor can appear all over the place in Fourier transforms and cause confusion. Some books prefer dealing not with ω but with frequency in Hertz (f) where $\omega = 2\pi f$, in this way all 2πs are in the exponent. In order to widen the scope of Fourier transform theory in the representation of sea waves (and waves in general) the power

$$\lim_{T \to \infty} \frac{1}{T} \int_{-T}^{T} [x(t)]^2 dt$$

rather than the energy is considered. The spectrum of this function is termed the "power" spectral density or just the spectral density and it is this that is usually dealt with by coastal and offshore engineers. This is as much detail as it is appropriate to delve into here. The subject of signal analysis and signal processing is a complex one and forms significant parts of electronics courses. The mathematical prerequisites are also beyond what this text has assumed.

Coastal engineers and marine physicists have expended considerable effort over the years to find the spectra that describe sea waves. After several oversimplified spectra, two have emerged that in general seem to fit observations. These are the Pierson-Moskowitz or PM spectrum and the JONSWAP (JOint North Sea WAve Project) spectrum. The waves on the surface of the sea are caused by the action of the wind. Precisely how this occurs remains one of the great unsolved mysteries (although there are several very well thought out contenders), nevertheless the consensus is that this is the case. If the wind has blown for long enough at a requisite speed and has enough sea area to act upon then the waves are neither duration limited nor fetch limited. Such a sea is termed fully developed and it is the spectrum of such a sea that is well described by the PM spectrum. The form of this spectrum is:

$$S_{PM} = 5 \left(\frac{H_s}{4} \right)^2 \frac{\omega_p^4}{\omega^5} \exp \left\{ -\frac{5}{4} \left(\frac{\omega}{\omega_p} \right)^{-4} \right\}$$

where ω is the frequency, ω_p is the frequency at the peak of the spectrum and H_s is the significant wave height (the average height of the highest one third of the waves). The shape of the spectrum $S_{PM}(\omega)$ is shown in Figure 7.13. If the sea is not fully developed, then the spectrum is a lot harder to determine. However, the importance of such a determination was enough for an expensive observational programme to be instigated in the early 1970s. The JONSWAP spectrum was the eventual outcome of these observations. The functional form of the JONSWAP spectrum is as follows

$$S_J = 3.29 \left(\frac{H_s}{4} \right)^2 \frac{\omega_p^4}{\omega^5} \exp \left\{ -\frac{5}{4} \left(\frac{\omega}{\omega_p} \right)^{-4} \right\} (3.3)^{\phi(\omega/\omega_p)}$$

where

$$\phi(x) = \exp \left\{ -\frac{1}{2\beta^2} (x - 1)^2 \right\}$$

$$\beta = 0.07 \qquad \omega \leq \omega_p \qquad \text{and} \qquad \beta = 0.09 \qquad \omega > \omega_p.$$

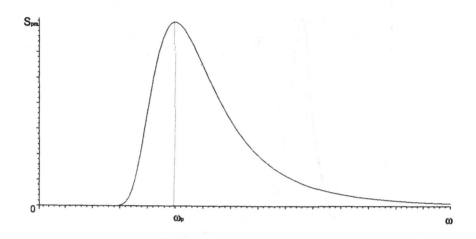

Figure 7.13: The Pierson-Moskowitz spectrum

The JONSWAP spectrum is not so widely accepted as the PM spectrum, especially outside Europe. The extra complications are to be expected given its applicability to fetch and duration limited seas, and the two values of β indicate a (slight) asymmetry. It is also more peaked than the PM spectrum, again to be expected as the fully developed sea will have had more time to spread its spectrum through various interactions. The shape of the JONSWAP spectrum is shown in Figure 7.14.

One aspect of real sea surface waves yet to be mentioned is directionality. The PM and JONSWAP spectra refer only to one dimensional waves whereas in reality sea surface waves are most definitely two dimensional. In engineering applications such as the forces due to waves on offshore floating or fixed structure this directionality is indeed very important and a great deal of research has taken place to model it. If waves are taken as two dimensional, then particular angular distribution functions have been proposed. A common choice is $|\cos \frac{1}{2}\theta|^s$ where θ runs from 0 to 2π and denotes the direction and s is a parameter dependent upon the frequency controlling the directional distribution of the wave energy. The expressions

$$s = s_p \left(\frac{\omega}{\omega_p}\right)^{-2.5} \qquad \omega \geq \omega_p$$

$$s = s_p \left(\frac{\omega}{\omega_p}\right)^{5} \qquad \omega \leq \omega_p$$

can be shown to fit a JONSWAP type of spectrum offshore. Here, ω_p remains the peak frequency and $s_p = 11.5(\omega_p)^{-2.5}$. More information about directional

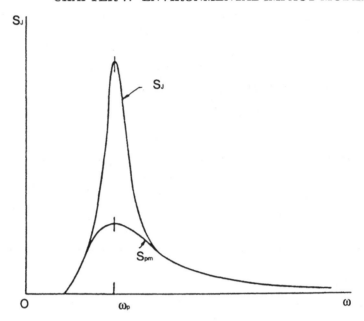

Figure 7.14: The JONSWAP spectrum

spectra can be found in the paper Mitsuyasu *et.al.* (1975). Fortunately, we are principally concerned with coastal processes where Snell's law of wave refraction means that the waves align themselves parallel to the coast. In fact it gets better, in applications to modelling coastal erosion, although knowledge of the PM and JONSWAP spectra can be useful as indeed is some knowledge of directional spectra, it turns out that whichever spectrum is used the predictions are similar! In attempting to formulate models to predict the movement of bed material, a commonly used technique is to find an "equivalent monochromatic wave" which turns out to be $\sqrt{2}U_{rms}$ where U_{rms} is the root mean square value of the spectrum (Soulsby (1997)). This is not true for very small values of mean speed, but these small speeds are of no concern here. The reason that U_{rms} features is that this quantity is the standard deviation of all the speeds that emerge from superposing all the orbits in a JONSWAP or PM spectrum. It is perhaps correct to be suspicious of linear theory and models based on linear theory when this is being applied to storm waves eroding a coast. There are alternatives, but each has its problems. The technique whereby linear theory is modified through expansion of variables in a suitable small dimensionless parameter (the wave slope is a popular choice) is called a perturbation technique and this has been successfully employed for over one hundred years. Military and civil defense shore protection codes are formulated using Stokes 2nd - 5th order solutions. This means that the variables in the governing non-linear equations have been expanded up to 5th power in the small parameter and like powers equated. This is used for water deeper than $0.01gT^2$ where T is the wave period. The wave has steep crests and shallow troughs when compared to sine waves (linear theory). For very shallow water (between $0.003gT^2$ to $0.016gT^2$) cnoidal theories have been used. Cnoidal wave theory is based on non linear equations and the surface wave profile is expressed in terms of Jacobi Elliptic functions which were never exactly common knowledge and are certainly rather obscure these days. Theories based on stream functions (similar range to cnoidal waves) together with other theories can be found in specialist books such as Sleath (1984) but none of them are without problems. The second order Stokes theory gives the following maximum speed of current under a crest due to a monochromatic wave U_c:

$$U_c = U_W \left[1 + \frac{3kh}{8\sinh^3(kh)} \frac{H}{h} \right]$$

whereas the maximum speed of current under a trough is U_{tr} where

$$U_{tr} = U_W \left[1 - \frac{3kh}{8\sinh^3(kh)} \frac{H}{h} \right].$$

The asymmetry is important and tends to drive sediment onshore (Soulsby (1997)). In these expressions, the quantity U_W is the wave orbital speed as dictated by linear wave theory:

$$U_W = \frac{\pi H}{T \sinh(kh)},$$

h is the water depth H is the wave height and $k = 2\pi/L$ is the wave number (L is the wave length).

7.3.4 Extreme Events

As good as the JONSWAP and Pierson-Moskowitz spectra are at wave prediction, they are at their worst when the situation is at all unusual. In offshore and coastal engineering, there has long been interest in being able to predict the "100 year wave". This is commonly thought to indicate the wave that only occurs every 100 years. This is wrong. The 100 year wave is that wave that has a low probability of returning with a frequency of less than 100 years. Precisely what level of probability to give depends on the statistical distribution assumed, and these statistical distributions are special to the statistics of extreme events. Exceptional storms, their associated floods and extremely high winds are the kind of phenomenon that environmental protection agencies are interested in predicting. They turn to the statistics of extremes to help them. Of course, a dynamic model based securely on hydrodynamics that in the natural course of running will predict the largest high water, the extreme current is ideal, but this approach has only really been successful in predicting storm surges. The flooding associated with storms, rivers spilling over their banks and the like still unfortunately surprise us. The major problem in predicting extreme events is the lack of data upon which to base predictions. Commonly, the kind of statistical distribution assumed contains a double exponential, that is the exponential the section on regression and fitting curves to data) is necessary in order to render the line of best fit a straight line. Unfortunately this bunches up almost all the data and renders extrapolation to home in on the extreme event subject to large uncertainty. In the theory of the statistics of extremes, there are many distributions from which to choose and the normal way of proceeding is to estimate parameters and allow these estimates to govern the choice of distribution. There are numerical examples in Chapter 8, and it turns out that the estimated parameters lead to the choice of the Gumbel distribution. For information here are some details for the Gumbel distribution:

$$p(x) = \frac{1}{\alpha} \exp\left\{ \left(\frac{x-k}{\alpha} \right) - e^{[(x-k)/\alpha]} \right\}$$

where α is a scale parameter of the distribution to be estimated. The second parameter k is a location parameter which again is estimated from data. Writing

$$F(x) = \exp\left(-e^{[(x-k)/\alpha]} \right)$$

and taking logarithms twice gives

$$\frac{x-k}{\alpha} = y = -\ln\{-\ln[F(x)]\}.$$

Plotting y against x thus gives a straight line, but the nested logarithm bunches up most data one would want to use to estimate α and k. Special "Gumbel" paper is produced to help with practical use. Another distribution perhaps more widely used in the civil engineering community is the Weibull distribution:

$$p(x) = \begin{cases} abx^{b-1} \exp\{-ax^b\} & x > 0 \\ 0 & x \leq 0. \end{cases}$$

The Gumbel and Weibull distributions come under the general class of General Extreme Value (GEV) distributions, but they all have the same problem: very long records are required for accurate estimation. It is the case that to predict with any certainty an event that has a return period of N years, a record of length at least $N/2$ is required! In how many sensitive locations are there even 25 years worth of wave data?!

On a practical level, engineers do what they can to design sea defenses and the like that can withstand the largest waves and strongest current likely to hit in say 100 years. The fact that the kind of stormy weather likely to give these extreme events can cause two "100 year waves" to arise in the same season is unfortunate and drives the non-technical to curse the modellers. The correct reaction is to call for more research; it is the case that all of the statistical techniques of this section, both the extreme statistics and the standard spectra are based upon the wave field being statistically stationary. This means it is assumed that the overall statistics over a very long time are unchanging. Global warming is precisely therefore what cannot be taken into account. All we can do is to suggest some over designing and hope that the extra cost is money well spent.

7.3.5 Interactions and wider issues

Here the effects of currents and waves have been examined separately. A good question to ask is whether there is interaction and if so what are their effects? Any interaction will be due to non-linearity, but this can take many forms. The interaction of a wave with itself produces currents and the Stokes' drift, the simplest form of this current, was derived in the previous chapter. The current due to a purely sinusoidal wave has an elliptic orbit, and if there is a current (for example a wind driven flow) in addition to this, then the interaction can enhance the net current to be greater than a simple summation of them would indicate. Of course, the interaction also causes Langmuir Circulation (see Chapter 6) which is a special helical current with distinctive important characteristics. Longshore currents can also be generated and these have their own special place in the consideration of coastal erosion. Waves being refracted by a sloping beach will give rise to a drift along the shore which is often responsible for the transportation of sand and like material. This is particularly important in beach replenishment studies. Another quite separate interaction is the modification of wavelength and wave celerity by the presence of a current. This modification can lead to different characteristics which may in turn cause increased erosion, as a longer wave will be more penetrating with depth as well as producing greater wave generated currents.

The all important interaction of course is that between the sea and the coast. Just exactly how does a storm induced wave and current tear into a vulnerable cliff? Of course we do not *exactly* know, but what is certain is that if the waves are prevented from slamming into the cliff or sea wall then the erosion is reduced. There is a considerable literature on the design of groynes and other coastal protection methods using physical barriers, and for further information the reader is referred to this which is clearly in the field of coastal engineering.

As far as managing is concerned the plan has to be as follows. First of all there is a pooling of knowledge supplemented by observation exercises where there are gaps. This knowledge must be recent, although it is important to know the historical development of particular coastlines. From this knowledge comes a recognition of those areas which are at risk from flooding or erosion, and cliffs that are unstable can be identified. All this takes place in a routine way, but how are the effects of global warming likely to be manifest? It is expected that there will be increased storm activity which leads to larger mean winds. These winds will in turn give rise to stronger wind driven currents as well as a more energetic sea surface. The JONSWAP spectrum will, it is expected, remain valid however the parameters within it will change to reflect this enhanced energy. The models of waves and currents given in this chapter and earlier can still therefore be used.

As there is more flooding due to increased rainfall as well as sea level rise, so the characteristics of groundwater are likely to change. This is a very important subject for land and river management but lies outside the scope of this text. As there is still some uncertainty about exactly how severe global warming will be it would be very expensive to guard against a "worst case" scenario which is the natural instinct of all civil engineers. The bolstering of coastal flood defenses alone would run into billions of dollars.

Before finishing, let us summarise some other consequences of global warming. There are other conservation issues such as the protection of sensitive salt marshes which house many rare and endangered species. There is the whole question of fisheries. Some species will thrive others will dwindle and the consequences for the industry which is particularly bedevilled with bureaucratic rules and regulations need to be thought through fairly. Water quality in terms of pollution has been dealt with elsewhere (Chapter 5) but issues such as the management of the quality of drinking water and the management of waste water although all very important are not tackled here. A discussion of all the consequences of global warming would fill several textbooks! We have glanced at those consequences that effect the coast.

Chapter 8

Modelling in Action

8.1 Introduction

This last chapter provides the reader with the opportunity to do actual numerical examples of marine modelling using the models that have been introduced in the first seven chapters. Do not worry if you have no expertise in mathematics, or if some of the more technical parts of the book so far have seemed impossible to understand, the whole point of this chapter is to start from the very beginning and to take you through simple models using a step by step approach. In writing this, the author has been very influenced by the books of the late K.A.Stroud, who has over the last 30 years published programmed learning texts in the UK for engineering students. These texts, though not the favourite recommended books by lecturers and teachers of engineering students, have proved extremely popular with students, particularly those who struggle with the technical aspects of mathematics. Therefore what I wish to achieve here is a similar easy to follow run through of some of the more elementary but nevertheless instructive marine modelling examples. As the problems are introduced, you are strongly advised to actually stop reading and do the problems *before* looking at the answer which will appear before the next part of the text. Although it is possible to give each problem a marine flavour, it turns out that in statistics in particular this often obscures the main point in that it makes what are quite simple principles seem complicated because of the nature of the details of the example. So although in what follows most of the examples are marine, this is not exclusively so.

8.2 Background Statistics

Statistics plays a very important role in marine science; one could even say a pivotal role. The principal difficulty in writing a text such as this is to cater for the wide variety of previous experience amongst the readership. The safest path to take is to assume very little previous knowledge. If what follows is too fast

paced then, certainly in the UK, there are a number of books that are designed
for students in the last two years of compulsory schooling to cover those parts of
statistics that appear in the National Curriculum. Generally, these have titles
that contain the phrase 'GCSE Mathematics' or 'Level 10 mathematics'. Some
of you of course will already be mathematically quite sophisticated, in which
case do pick and chose from what follows. We shall start with the revision
of what statisticians call *measures of central tendency*, which means ways of
assessing where the middle of a set of data is. The simplest form of data is a
list of numbers, although data are also often produced in the form of frequency
tables. We shall deal with both.

Example 8.1 *We wish to find the mode, median and mean of the following list
of numbers:*

$$5, 3, 6, 5, 4, 5, 2, 8, 6, 5, 4, 8, 3, 4, 5, 4, 8, 2, 5, 4.$$

Solution
First of all, do not worry about the definitions of these words; instead, we put
the numbers in ascending order as follows:

$$2, 2, 3, 3, 4, 4, 4, 4, 4, 5, 5, 5, 5, 5, 5, 6, 6, 8, 8, 8.$$

The mode is the number that appears the most times.

   mode =

The median is the number which is in the middle of the distribution:

   median =

Finally, the mean of the numbers is the sum of the numbers divided by 20
(there are 20 numbers in all):

   mean =

You should have obtained the answers: mode = 5, median = 5, mean = 4.8.

In this example, there is a clear mode since there are six 5's, and fewer of
each of the other numbers (in general there is often a tie). There is an even
number (20) of numbers, therefore the median is the average of the tenth and
eleventh numbers. Since both of these are 5, so is the median. The mean is,
uniquely, 4.8. Next let us consider something a little more usual in scientific
applications, that is, a situation where the numbers are grouped into classes and
we have in effect a frequency distribution. This is usually given in tabular form.
Table 8.1 gives the numbers of zooplankton of various lengths as measured by a
marine biologist (adapted from research data and considerably simplified). The
frequency polygon associated with these data is shown in Figure 8.1.

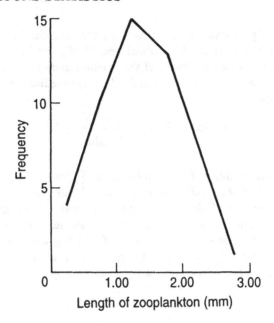

Figure 8.1: A frequency polygon

Length of zooplankton (mm)	Number of zooplankton
0.01-0.50	4
0.51-1.00	10
1.01-1.50	15
1.51-2.00	13
2.01-2.50	7
2.51-3.00	1

Table 8.1 *A Frequency Table*

The median of these data is still the middle number, but this is troublesome of find when the data take this form. In fact it is the value taken by the horizontal scale when a vertical line precisely divides the area under the frequency polygon into two equal halves. The mode is the peak of the frequency polygon. The mean is the quantity μ, which is given by the formula:

$$\mu = \frac{\sum f_i n_i}{\sum n_i},$$

where the letters f and n denote the frequency of occurrence of the number, and the number, respectively. The subscript is there to designate that there are many numbers (i would run from one to six in our example) and the $\sum$ sign denotes that summation over all i is to occur. The mean is always uniquely defined, although the same cannot be said for either the median or the mode.

The mode is, straightforwardly, the class that contains the largest number, but the median either has to be determined graphically, or by a rather complex formula derived from its definition. If the median occurs in a particular class, and the lower boundary of this class is L, then the median itself is determined from the formula:

$$\text{Median} = L + \left(\frac{N/2 - (\sum f)}{f_{\text{median}}} \right) c,$$

where N is the total number of items in the data, $\sum f$ is the sum of frequencies of all classes *lower* than the median class, f_{median} is the frequency of the median class, and c is the size of the median class interval. Given grouped data, it is easy enough to spot in which class the median lies; all the above formula represents is a mathematically precise way of dividing the area of this class to ensure that the median line so derived cuts the total area under the frequency polygon precisely in half. Now have a try at calculating the mode median and mean:

mode =

median =

mean =

You should have obtained the answers: mode = 1.255 mm, mean = 1.375 mm and median = 1.3969 mm. In this problem we meet several features that are typical in the handling of data. The mode is simply the mid-point of the interval (1.01-1.50) that contains the greatest number of animals. The mean follows by applying the formula remembering that in this instance the number of animals is multiplied by the length of zooplankton corresponding to the *middle* of the range (for example, 4×0.255 is the first entry in the numerator, 10×0.755 is the second, etc.). Finally, the median is calculated using the given formula with $L = 1.01$, $c = 0.49$, $\sum f = 14$ and $f_{\text{median}} = 15$. There are some more data sets that you can practice on at the of this chapter.

Of course, in a brief summary such as this it is not possible to go into much detail in the way of statistical theory, nor is it desirable. The many specialist texts on statistics, having started as we have by introducing measures of central tendency, go on to discuss topics such as standard deviation, distributions, probability, and then to applied topics which include sampling, regression, hypothesis testing and experimental design. All of these have a role to play in marine science, but it would be over-ambitious to try to cover them in this book. Perhaps the most important point to make is that the central purpose of statistics is *inference*. The reason why data are analysed is to enable scientists to establish hypotheses (in a statistical sense) from the data. For this reason, the more theoretical aspects of probability theory are omitted here with the view that, should any be required it can be introduced *in situ*, as it were. The specialist texts are there for those hungry for more theory.

However, we do need to define variance and standard deviation. The variance of a set of numbers measures how spread out they are from their mean. It is defined by the formula:

$$\sigma^2 = \frac{\sum(X_i - \overline{X})^2}{N},$$

where the symbols have the following meanings: X_i denotes the data (i.e. the numbers themselves), $\overline{X}$ is the arithmetic mean, $\sum$ is the summation sign which means that each number has the mean subtracted from it before it is squared, then the whole is divided by N, the number of numbers in the data set. The reason behind squaring each difference is that this makes all entries under the summation sign positive, hence making sure that the result of this sum is indeed a true representation of the spread of the data from the mean. Statisticians call this a 'measure of dispersion', but this is not an appropriate expression to use in a marine modelling book! In order to restore the dimensions, the variance is normally square rooted (hence the square on the left-hand side) and the symbol σ is called the standard deviation.

Here is a practice example.

Example 8.2 *Find the variance and standard deviation of the numbers:*

$$5, 3, 6, 5, 4, 5, 2, 8, 6, 5, 4, 8, 3, 4, 5, 4, 8, 2, 5, 4.$$

Solution

Try calculating the two quantities:

variance =
standard deviation =

You should have obtained the answers $\sigma^2 = 3.116$ and $\sigma = 1.765$. If you 'cheated' and used a calculator or a computer, this is no problem as long as you are sure of what you have calculated and know what standard deviation and variance actually mean. The above answers only validate your arithmetic; they do not confirm your understanding! When a frequency table is involved, the definitions are of course the same but the method of calculation looks a little different. In fact, there is a very useful formula that can be derived from the definition of variance that proves useful in calculation. This states that the variance is given by the expression:

$$\sigma^2 = \overline{X^2} - \overline{X}^2,$$

which can be read as variance equals the 'mean of the squares minus the square of the mean'. For grouped data, the following expression is the formula for variance:

$$\sigma^2 = \frac{\sum fn_i^2}{N} - \left(\frac{\sum fn_i}{N}\right)^2;$$

the standard deviation is of course the positive square root of the variance. Determine the variance and standard deviation of the data presented in Table 8.1:

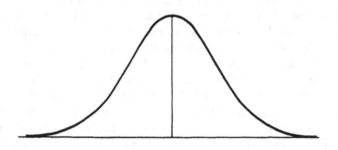

Figure 8.2: The normal distribution

variance =
standard deviation =
You should have found that the variance is $0.408mm^2$ and that the standard
deviation is $0.639mm$. Once again, the use of a calculator or software on a mi-
crocomputer eases the mathematical chore, but this must not be at the expense
of understanding. There are more examples upon which to practice at the end
of the chapter. Let us now turn to applied statistics.

We have not dwelt on probability or distribution theory, since these topics
are not central to our purpose, but once the mean and standard deviation are
fixed, it is often acceptable to assume that data presented in the form of a
frequency distribution approximate closely to normal. This assumption tends
to be universal. In fact it tends to be assumed even when it is not appropriate;
users of statistical routines need to be aware of this. The first thing to remember
is that the normal distribution assumes that the variation of frequency f with
random variable x is the function:

$$f(x) = \frac{1}{\sqrt{2\pi}\sigma} \exp(-(x-\mu)^2/2\sigma^2) \quad -\infty < x < \infty,$$

where σ is the standard deviation and μ is the mean. The general shape of
this curve is shown in Figure 8.2. It is obviously desirable to have only a single
distribution to cater for predictions based on the normal distribution. This has
the consequence that all tables and software that contain information about the
normal distribution have adjusted to a mean of zero and a standard deviation
of one. Remember therefore to transform your data X into the normal variable

z, sometimes called the z-statistic, through the simple transformation:

$$z = \frac{X - \mu}{\sigma},$$

before doing any statistical testing. The commonest of tests to use is the χ^2 test, which can be used to test whether or not a particular set of data fits a given hypothesis. A simple example is the toss of a coin. If a given coin is tossed 1000 times, say, and the outcomes are recorded, then this test can be used to decide whether the coin is biased or fair. Similarly, the χ^2 test can be used to decide whether or not data fit the conclusions drawn from a particular model. As hinted at above, however, one never gets *the* answer, and the criterion for acceptance or rejection of a hypothesis, to the applied marine scientist, is not God-given but is in fact dependent on assumptions involving the normal distribution.

Before we can do examples, we need to introduce the subject of hypothesis testing. This is the traditional first step on the road to *inference*, the main purpose behind most of statistics. Suppose we have some data, perhaps from observations taken on an oceanic cruise. These data form what statisticians call a population. It is a collection of numbers arranged in a table or represented graphically. There will be certain statistics associated with the data - we have calculated the mean and standard deviation, but there are others. Now suppose further that we suspect that these data obey the form dictated by, say, the normal distribution. That is, we suspect that the mean and standard deviation conform to a certain normal, bell-shaped curve. We can use the χ^2 test to ascertain the truth of this hypothesis. This hypothesis is called the *null hypothesis* and is given the symbol H_0. If H_0 is rejected when in fact it is true, we say that a type I error has occurred. If we accept H_0 when it is actually false, we say that a type II error has occurred. Unfortunately, it is all too easy to make both sorts of errors, and it is always best to take a cynical look at the data, looking for oddities (*outliers* as statisticians call them) which may be due to human error in observing, or instrumental failure, and which could distort the data and be the underlying cause of the type I or type II error. Finally, statisticians give the symbol H_1 to an alternative to the null hypothesis. Hopefully some of this will come alive through the next two examples.

The first of these examples is an introductory one involving that old standby, the tossing of coins; the second is a more practical example involving real marine data.

Example 8.3 *Suppose a coin is tossed 1000 times, and the outcome is 530 heads and 470 tails. We might expect the outcome to be 500 heads and 500 tails, but then again it is the nature of chance that most of us would actually be surprised at such a precise obedience of the laws of probability. The pertinent question to ask is: is the coin fair? In other words, can the deviation from the ideal answer be attributed to chance, or is there a bias in the coin? In this case, the null hypothesis might be:*
H_0: *heads and tails occur with equal frequency.*

Solution

We shall use the χ^2 test. In order to do this, we need an appropriate distribution. The χ^2 distribution can be found in Appendix A. In this table, the top row, labelled χ^2 which denote the *levels of significance*, gives a choice of thirteen numbers. These numbers represent significance levels so that, respectively, the columns that they head are appropriate to testing at the 99.5%, 99%, 97.5%, 95%, 90%, 75%, 50%, 25%, 10%, 5%, 2.5%, 1% and 0.5% levels. Let us choose the value 0.01, so that we are testing at the 99% significance level. Coin tossing is a process that has two possible outcomes (heads or tails); therefore the first row is chosen. The number in this row is

$$\chi^2 =$$

You should have read the number 6.635. Now we calculate the value of χ^2 according to the formula:

$$\chi^2 = \sum \frac{(\text{Observed} - \text{Expected})^2}{\text{Expected}}.$$

Remember, the summation sign is not a sum over 1000 trials, but a sum over all possible outcomes. The calculated value is:

$\chi^2 (\text{calculated}) =$

The calculation should have proceeded as follows:

$$\chi^2 = \frac{(530 - 500)^2}{500} + \frac{(470 - 500)^2}{500},$$

so that $\chi^2 = 3.6$. This value is less than the value in the table, so we accept the null hypothesis H_0 and conclude that, at 99% significance level, the coin is not biased. This is probably the correct conclusion, but if on examining the data we found 200 consecutive heads, we would want to research further into how the coin was tossed, etc. This latter point may seem a little silly here, but if we were dealing with real data, it is analogous to 'eye-balling' the figures and spotting if anything suspicious is present in the data. Mind you, one is much more likely to look if the hypothesis is rejected.

To the relief of most of you, the next part comprises a few marine related examples.

Here is an example involving fish. Table 8.2 gives the actual and expected values for catches of five species of fish.

	Species A	Species B	Species C	Species D	Species E
Expected catch	25	5	7	31	35
Actual catch	20	4	17	26	30

Table 8.2 *Actual and Expected catches of five species of fish*

First, state the null hypothesis for this problem:

H_0 states that:

You should have written 'H_0 states that the expected catch and the actual catch are the same'. Using a χ^2 test with parameter 0.01, do we reject H_0? To answer this we calculate χ^2 from the formula and get the appropriate value of χ^2 from the table in Appendix A:

χ^2(calculated) $=$

χ^2(table) $=$

From your calculation and from the table, you should have obtained the values

χ^2(calculated) $= 17$,

χ^2(table) $= 13.3$.

On the face of it, these results indicate that we should reject the null hypothesis. However, if we glance at the table of data, there is a very large discrepancy between expected and actual catch for species C. Without species C data, H_0 would have easily been accepted. The correct conclusion to draw therefore is that the figures for species C need to be re-examined and the reason for the glut of fish or the serious under-estimation of the catch ascertained. In passing, note that for an n-variable problem ($n=2$ for the coin, and $n=5$ for the fish) we look at the line $n - 1$ rather than line n in the χ^2 table in Appendix A. The reasons for this are rather technical and have to do with the (statistical) degrees of freedom of the system.

When using hypothesis testing, it is always necessary to put data in the form of frequency. This was alluded to in Chapter 4. The χ^2 test simply does not work for dimensional data in the form of lengths or masses. If your data is in such a form, classify them in some way, batch them up to rid the numbers of dimension. Once the data is put into these m classes, then the degrees of freedom, the row along which to look up the value in the χ^2 table is $\nu = m - k - 1$ where k is the number of parameters to be estimated (usually zero for us). Let us go no further here; interested readers are directed to statistics books - there are plenty to choose from. (The statistics chapters in Modern Engineering Mathematics and Advanced Modern Engineering Mathematics by Glyn James, published by Pearson, are particularly accessible and to my taste.)

The final topic to cover in this briefest of excursions into statistics is fitting lines to data. The most common example of this is the regression line, which is a line of best fit through a set of data points.

Given a scatter plot as shown in Figure 8.3, there is a quite straightforward procedure for drawing a line of best fit through the data. An arbitrary line is drawn, then the square of the perpendicular distance of each point from this line is calculated. These are all added together, and the minimum value of this is found. The parameters of the line that correspond to this minimum value give the line of best fit. Difficulties arise only when the data are so scattered that there is virtually zero correlation, in which case the line of best fit has no meaning. In fact, there are always *two* regression lines. If x and y denote the standard axes, these regression lines are called 'y on x' and 'x on y', and if there is no correlation then these two regression lines are at right angles to each other. Recall that the term correlation refers to the measure of agreement between two sets of data. A correlation of 1 denotes perfect agreement, a correlation of -1 denotes perfect disagreement (as an example of the latter, the rainfall at one point of an estuary, and the salinity of the water at the same point; as the

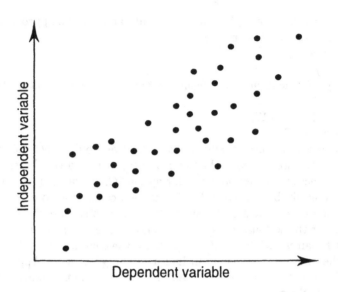

Figure 8.3: A scatter diagram

rainfall increases, the salinity decreases and vice versa) and a correlation of zero denotes no relationship at all. Other complications occur when there is obviously a relationship between two variables, but this relationship is not a linear one. This takes us into the heady topics of log-linear and log-log plots. These days most schools seem to have abandoned logarithms because they are no longer of any practical use as a calculating tool. They have gone the way of the *ready reckoner* and the *slide rule*. Unfortunately, logarithms have another function that has not and is never likely to be superseded - they are used to represent data where some kind of exponential growth is taking place. Those students that need to know about such things, such as students of biology and marine science, are thus faced with logarithms for the first time. The good news is that fortunately, there is no need to dwell at length on the many properties of logarithms, just a few; all that is necessary in fact is given below.

If an animal is growing exponentially, then its weight w might be related to time t through a relationship such as:

$$w = a + b \exp(ct)$$

where a, b and c are known constants that are fixed once the species and its environment are also fixed. In reality of course this growth will stop, but this is a simple illustration only. If we wanted to make t the subject of this formula, then we would subtract a from both sides before taking logs to obtain:

$$t = \frac{1}{c} \ln\left(\frac{w - a}{b}\right),$$

where the symbol 'ln' denotes the natural or Naperian logarithm. This particular logarithm function is the inverse of the exponential function, and is the 'log' referred to in the phrase 'log-linear' as in graph paper. We have still not given the reasons for needing to know about such graph paper. To do so, consider the expression just derived,

$$t = \frac{1}{c} \ln \left(\frac{w - a}{b} \right).$$

If data corresponding to $(w - a)/b$ were to be plotted on one axis of log-linear paper, and data corresponding to t be plotted on the other, then provided w and t were related in the way dictated by the above equation, the plot would be a straight line (with slope c). Once a scatter plot can be assumed to contain within it an implied linear relationship, then all the regression methods developed for straight lines can be brought to bear on the data. Table 8.3 gives some examples of relationships and the paper that should be used to display them as a straight line. In what follows, X and Y are the independent and dependent variables, respectively.

Equation	Straight line	Description
$Y = \frac{1}{a+bX}$	$\frac{1}{Y} = a + bX$	A hyperbola: use ordinary graph paper
$Y = ab^X$	$\ln Y = \ln a + X \ln b$	An exponential curve: use log-linear graph paper
$Y = aX^b$	$\ln Y = \ln a + b \ln X$	Geometric curve: use log-log graph paper
$Y = \frac{1}{ab^X + c}$	$\frac{1}{Y} = ab^X + c$	Logistic curve: use log-linear graph paper (with care!)

Table 8.3 *Some common relationships*

It is possible to obtain commercially special paper that renders variables that are related logistically (the last entry in Table 8.3) as a straight line. However, log-linear paper can be used provided the equation is transformed into exponential type by treating $(1/Y) - c$ as a variable.

One important question we have not yet addressed is how to assess whether or not a particular law is suitable for a given set of data; we cannot always rely on simply 'eye-balling' it. If we wish to compare two sets of figures in a quantitative manner, then we calculate a correlation coefficient. There are several such coefficients to choose from, but the one most commonly used is the Pearson correlation coefficient, which is 1 for perfect agreement, -1 for perfect disagreement, and 0 for no relationship at all. To calculate the Pearson correlation coefficient, r_{XY}, the formula:

$$r_{XY} = \frac{(1/N) \sum [(X_i - \overline{X})(Y_i - \overline{Y})]}{S_X S_Y},$$

where

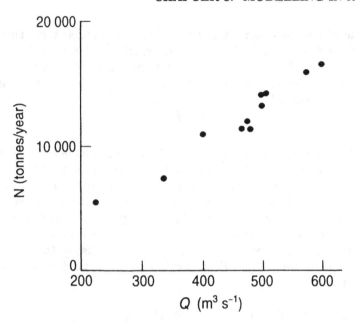

Figure 8.4: Total nitrogen against river flow

$$S_X^2 = \frac{1}{N-1}\sum(X_i - \overline{X})^2 \text{ and } S_Y^2 = \frac{1}{N-1}\sum(Y_i - \overline{Y})^2,$$

is used. All summations are over all the data points. The presence of $N-1$ rather than N in some of these expressions may perplex some readers, but it is a consequence of various statistical assumptions, in particular the assumption that the variates when standardised to a mean of zero and a standard variation of one obey a slightly distorted normal curve called a *t-distribution*. Although the above formula gives the definition of r_{XY}, we give below the most widely used practical formulae for calculating not only r_{XY} but also the regression line of Y on X in the form $Y = AX + B$. Purists will notice a missing factor of $(N-1)^2/N^2$ in the formula for r_{XY}, but this quantity is very close to one in most practical examples. In fact, if it is not, then any straight line drawn through such sparse data has only scant value.

$$r_{XY}^2 = \frac{(N\sum XY - \sum X \sum Y)^2}{|N\sum X^2 - (\sum X)^2||N\sum Y^2 - (\sum Y)^2|},$$

$$B = \frac{N\sum XY - \sum X \sum Y}{N\sum X^2 - (\sum X)^2}, \quad A = \frac{\sum Y - B\sum X}{N}.$$

Let us now do an example.

Example 8.4 *Table 8.4 gives the discharges of nitrogen (N) and total phosphorus (P) through the River Göta in tonnes per year, as measured in the years 1972-82 (inclusive). the quantity Q denotes the river discharge in $m^3 s^{-1}$*

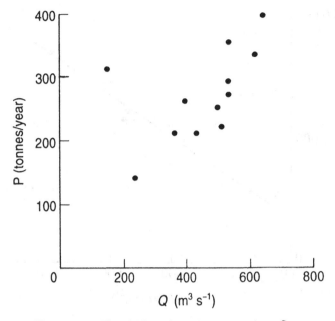

Figure 8.5: Total phosphorus against river flow.

Discharge	1972	1973	1974	1975	1976
N(t/yr)	13600	8900	13500	12700	7000
P(t/yr)	310	210	250	220	120
$Q(m^3s^{-1})$	150	365	505	515	240

Discharge	1977	1978	1979	1980	1981	1982
N(t/yr)	16700	14900	13600	18700	18000	16400
P(t/yr)	350	290	310	390	330	270
$Q(m^3s^{-1})$	535	535	435	645	620	535

Table 8.4 *A Table of river discharge data*

Solution

First, we need to plot the two scatter diagrams of the discharges of nitrogen and phosphorus. These are shown in Figures 8.4 and 8.5, respectively. The variable Q is the independent variable, and it is seen that the data are suitable for a linear regression line to be appropriate. Calculate the two correlation coefficients r_N and r_P:

$r_N =$ $r_P =$

Whether you used the formula directly, used a calculator or software, you should have obtained the answers $r_N = 0.73$, $r_P = 0.53$. Although both correlations are positive, they are not particularly high, so it is not obvious that linear regression is the best way to obtain reliable predictions. One may find a better non-linear relationship, but looking at scatter plots does not immediately

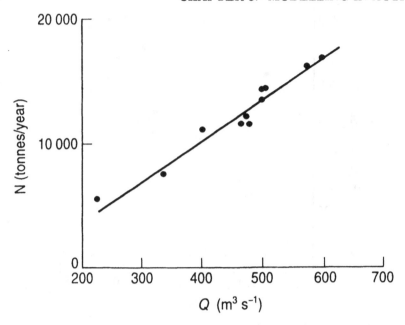

Figure 8.6: Regression line for nitrogen.

suggest any obvious alternative candidates. We therefore still press ahead and calculate the linear regression lines, but bearing in mind that predictions need to be treated with some caution. It is possible in fact to place error bars on the values of r_{XY}, but such refinements are considered outside the scope of this introductory text. Try calculating the two regression lines for these data using the formulae:

For nitrogen: $N = A_N Q + B_N$

$A_N =$

$B_N =$

For phosphorus: $P = A_P Q + B_P$

$A_P =$

$B_P =$

You should have obtained the answers:

$$A_N = 17.06, \quad B_N = 6121.06, \quad A_P = 0.258, \quad B_P = 158.22.$$

Note that no attention has been paid to the units here. The data are given in a mixture of units, as is quite typical (one might even say prevalent) in marine science with its long nautical traditions, and no conversions to, say, standard SI units have been made. This may annoy the purists, but in the calculation of lines of best fit, the geometric distance between the data points and the regression line has been minimised and this process is independent of units. Only if we wish for sensible units for these constants A_N, A_P, B_N and B_P does it become necessary to standardise. Finally in this example, let us do some predicting. Use the regression lines to predict the values of nitrogen and phosphorus in the

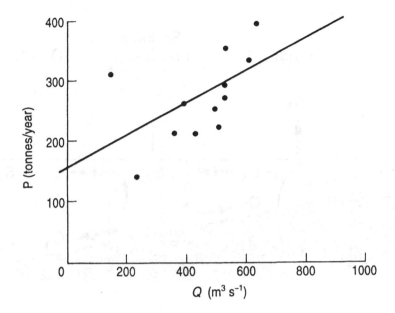

Figure 8.7: Regression line for phosphorus.

River Göta when the river discharge is 800 $m^3 s^{-1}$.

$N =$

$P =$

Either from drawing these lines on the graphs (shown in Figures 8.6 and 8.7) or, more accurately, from inserting $Q = 800$ into the formula for each line in turn, you should have obtained the answers:

$$N = 19769, \quad P = 364.$$

The first albeit less accurate method is acceptable, particularly because it keeps you in touch with the data, reminding you how scattered the points are, and hence how low the correlation is. Most importantly, it indicates how much (little?) reliance can be put on these predictions. A correlation of 0.9 would be considered a reasonable figure, and the data falls well below this.

This is as far as we will go into using statistics for modelling. The next step would be to look at non-linear regression and to include placing confidence intervals on predictions. Those interested in these topics need to consult more specialist statistics texts. The subject of statistics does return when we consider wave prediction a little later in this chapter.

Let us now turn to our attention to modelling how the sea moves. We shall utilise our knowledge of dimensional analysis, simple dynamic balances as well as some numerical methods.

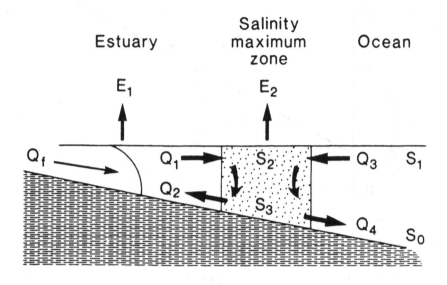

Figure 8.8: The internal estuarine circulation

8.3 Modelling in Action

In this section, we shall try to do some modelling. This is the longest section in this book, but is arguably also the most important. Please be prepared to do some calculations. These will progress from involving only simple algebra and arithmetic through to examples involving calculus. It is possible to get some way by simply obeying rules, but eventually you will need to understand what you are doing. This first example involves simple balances in an estuary.

Example 8.5 *Figure 8.8 shows the circulation of an idealised estuary. The river flux is Q_f and the internal fluxes are labelled Q_1, Q_2, Q_3 and Q_4. If $Q_f = 10m^3s^{-1}$ and the following salinity and evaporation values have been measured $S_0 = 29.5p.p.t.$, $S_1 = 29.3p.p.t.$, $S_2 = 35.6p.p.t.$, $S_3 = 35.8p.p.t.$, $E_1 = 2m^3s^{-1}$ and $E_2 = 250m^3s^{-1}$, (p.p.t. signifies part per thousand). Calculate the internal fluxes from conservation laws. [Adapted from Wolanski (1988)]*

Solution For this problem, we simply observe that both salt and water are conserved. It is a box model. For the estuary to the left of the maximum salinity zone, the conservation of water means that we must have

$$Q_f + Q_2 = Q_1 + E_1.$$

Salt is neither created nor destroyed in the maximum salinity zone, therefore

$$Q_1 S_2 = Q_2 S_3.$$

Considering the overall balance in the estuary, the conservation of water gives

$$Q_f + Q_3 = Q_4 + E_1 + E_2,$$

and the salt balance in the mouth gives

$$Q_3 S_1 = Q_4 S_0.$$

These equations can be solved (four equations in four unknowns). In particular

$$Q_3 = \frac{S_0(E_1 + E_2 - Q_f)}{S_0 - S_1}$$

and

$$Q_1 = \frac{S_3(Q_f - E_1)}{S_3 - S_2}.$$

Putting in the figures given yields the results

$$Q_1 = 1.43 \times 10^3 m^3 s^{-1}, \quad Q_2 = 1.42 \times 10^3 m^3 s^{-1},$$

$$Q_3 = 3.57 \times 10^4 m^3 s^{-1}, \quad \text{and} \quad Q_4 = 3.54 \times 10^4 m^3 s^{-1}.$$

In Chapter 2 we met dimensional analysis. This is a device that tells us which terms (i.e. processes) are important and which can be neglected in a given oceanographic situation. The first example involves the Gulf Stream.

Example 8.6 *An ocean current of 1 ms^{-1} is 200 km wide. It moves in a path with a radius of curvature 1000 km. Calculate a Rossby number if the latitude is 40°N. Discuss the modelling implications but ignore friction entirely.*

Solution
The first step is to write down dimensions that are appropriate to speed U, length L, and Coriolis parameter f, since these make up the Rossby number $Ro = U/fL$. From the question statement, we can estimate these parameters:

$$U = \qquad L = \qquad f =$$

Deciding upon a typical speed is not difficult since $U = 1 ms^{-1}$ is given in the question, and so are two typical lengths. $L_1 = 2 \times 10^5$ m is the cross-stream dimension, and $L_2 = 10^6$ m (= radius times one radian of angular measure) is a measure of length associated with the curvature of the Gulf Stream itself. The Coriolis parameter $f = 2\Omega \sin(latitude)$, where $\Omega = 7.29 \times 10^{-5} rads^{-1}$ is the angular speed of the Earth is not controversial. It gives $f = 10^{-4} s^{-1}$. From these figures, the two possible values for the Rossby number are:

$$Ro_1 = \qquad Ro_2 =$$

Using the two possible values of L, the two values of Ro are 0.05 and 0.01, both of which are small enough for the advective terms in the equations of motion to be ignored. In fact, the Gulf Stream can be analysed using a linear theory unless bends in the stream or the narrowness of the stream itself lead us to choose $L = 10^4$ m, in which case the model has to be non-linear. This is entirely consistent with present theories in which non-linear terms are only important when relatively small-scale features are coupled with high current speeds. The power of dimensional analysis is clearly demonstrated, but also its principal weakness in that we now have to actually *solve* the equations, which dimensional analysis cannot help us to do. The models of Stommel and Munk, mentioned in Chapter 2 were developed after such dimensional analysis.

Example 8.7 *In this example, let us consider a coastal sea where friction is important. There is a tidal current of magnitude 1 m s^{-1}. The sea level rise 1 m and appreciable differences in the horizontal occur over distances of 100 km. The latitude is 56° N, and we assume a quadratic friction law of the form:*

$$friction = \rho C_D \mathbf{u}|\mathbf{u}|,$$

where ρ is the density , C_D is a drag coefficient of magnitude $\sim 10^{-4} m^2$, and $\mathbf{u}$ is the fluid velocity just above the sea bed. Determine the magnitudes of the terms in the equation of motion,

$$\frac{\partial u}{\partial t} + u \frac{\partial u}{\partial x} - fv = -g \frac{\partial \eta}{\partial x} + \frac{friction}{\rho},$$

and determine what the magnitude of the horizontal gradients have to be in order for the term $u(\partial u / \partial x)$ to become important.

Solution
For this problem, typical dimensions are:

$$U = \qquad L = \qquad T =$$

We are given directly that $U = 1$ m s^{-1}, but what is L? In the vertical, a typical length might be 1 m, but this model does not include vertical dynamics (no variation with z, the vertical component). So we deduce that L must represent a typical horizontal length scale which, from the statement of the question is 10 km or 10^5 m (compatible units are important here). Assigning T has the potential for being awkward since there are two naturally occurring time scales: one is the Coriolis parameter, and the other is the frequency of the tide. In mid-latitudes, the Coriolis parameter is $\sim 10^{-4} s^{-1}$, and the frequency of the tide, corresponding to the M_2 semi-diurnal tide is:

$$f_{M_2} = \frac{\pi}{day} = 1.4 \times 10^{-4} s^{-1}.$$

So, fortuitously, T is of magnitude $10^4 s$ satisfies both time scales. It is relevant to ask what would have been done if the time scales had been different. Which one would we have chosen? The answer is the tide, since it is the dynamics of tides that interests us in this problem (in fact, the Coriolis acceleration ceases to be an important factor if the tidal acceleration has a much larger magnitude as might be the case, for example, in a narrow river; this would emerge from the dimensional analysis). Pooling this information, we determine the magnitudes of all the terms in the equation. We find that:

$$\frac{\partial u}{\partial t} = \qquad u\frac{\partial u}{\partial x} =$$

$$fv = \qquad g\frac{\partial \eta}{\partial x} =$$

$$\frac{\text{friction}}{\rho} =$$

Using $U = 1$, $T = 10^4$ and $L = 10^5$ you will find that *all* of the above terms are of magnitude 10^{-4} except the $u(\partial u/\partial x)$ term (the non-linear term), which is 10^{-5}. Remember that $g = 10$ (approximately) so the magnitude of the term $g(\partial \eta/\partial x)$ is also 10^{-4}, bearing in mind that the tidal elevation is of the order of 1 m. Hence, only this non-linear term may be discarded from our tidal model, and even this is not clear. For example, it cannot if $L \approx 10^4$ m (=10 km), that is, there are features on the 10 km scale (embayments, headlands, inlets). Again it is recognised that it is important to take non-linear processes into account near coasts where such features predominate. Thus in this example, dimensional analysis only confirms that it is really not safe to omit any terms and a numerical method must be used to solve the equations in their entirety. Here is another example of the use of dimensional analysis but this time differences are also calculated. In this example, it is assumed that you have no experience of calculus whatsoever, for those that have the language may be too elementary, even a little insulting. If it is, skip it (although glance at it first, you still might learn something!)

Example 8.8 *A cool sea is adjacent to a warm current, and temperature θ diffuses according to the diffusion equation:*

$$\frac{\partial \theta}{\partial t} = \kappa \frac{\partial^2 \theta}{\partial x^2},$$

where the partial derivative symbols are (of course) gradients. $\partial\theta/\partial t$ means the rate of change of temperature with time, and $\partial^2\theta/\partial x^2$ is a rate of change of the temperature gradient $(\partial\theta/\partial x)$ with respect to distance(x). Typical scales will be given, since the purpose of this particular example is to run through numerical approximation rather than to give another example of dimensional analysis. Typical lengths are 100 km, times are consistent with the Coriolis parameter, and κ, the diffusion coefficient, has magnitude $4 \times 10^5 m^2 s^{-1}$. Figure 8.9 gives the locations of various measurement sites marked A, B and C. At site A, the gradient of the temperature is $1°C$ in 50 km; at site B, which is 100 km east of A, the gradient is $1°C$ in 40 km. The directions of these gradients are indicated in Figure 8.9 by the arrows; they are all positive. At site C the temperature is $10°C$. Use the information given to estimate what the temperature will be in 24 hours time.

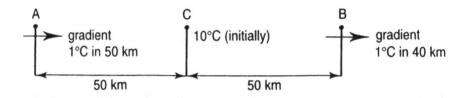

Figure 8.9: Calculating differences from the data.

Solution

The first task is to check that the dimensions of both sides of the diffusion equation are the same and that their magnitude is of the same order. Temperature cannot be expressed in terms of M, L and T (mass, length and time), but since it occurs once on both sides of the diffusion equation, this part at least is in balance. Next, we write down the dimensions of the gradients of θ:

$$\frac{\partial \theta}{\partial t} \text{ has dimensions} = \qquad \frac{\partial^2 \theta}{\partial x^2} \text{ has dimensions} =$$

Your answers should be θT^{-1} and θL^{-2}, respectively. Each time a gradient is required, a derivative is taken which is equivalent to a division by the quantity that varies (the independent variable). Hence, $\kappa(\partial^2 \theta/\partial x^2)$ has dimensions $L^2 T^{-1}.\theta L^{-2} = \theta T^{-1}$, which is the same as the dimensions of $\partial \theta/\partial t$. We are given the dimensions of L and T; these are $L = 10^5$m and $T = 10^{-4}$s, and since $\kappa = 4 \times 10^5 m^2 s^{-1}$, we conclude that the two sides of the diffusion equation have the following magnitudes:

$$\frac{\partial \theta}{\partial t} \approx \qquad \kappa \frac{\partial^2 \theta}{\partial x^2} \approx$$

The results of each calculation are $\theta \times 10^{-4}$ for the left hand side, and $4\theta \times 10^{-4}$ for the right hand side. This is acceptable bearing in mind that we are only considering orders of magnitude; i.e. we are not saying that 1 is equal to 4, but that 1 and 4 are both in the same ball-park (between 1 and 10, for example). The next part of the question moves us away from such ball-park calculations and forces us to be a little more precise.

The unknown in the last part of the question is the left hand side, since it is this that will tell us how the temperature changes with time. There is enough information to calculate the right hand side as follows. We are given two spatial gradients, and we can use these to calculate how the gradient itself is changing, which gives an approximation to the $\partial^2\theta/\partial x^2$ term. However, since we are undertaking actual computations of quantities here, units are important and we need to convert to metres. Doing this, the gradients are as follows:

$$\text{At A} \qquad \text{At B}$$

The answers are, at A, $\frac{1}{5} \times 10^{-4}\theta°C\ m^{-1}$ and at B, $\frac{1}{4} \times 10^{-4}\theta°C\ m^{-1}$. The gradient of the gradient of θ at the mid-way between A and B is therefore as estimate of $\partial^2\theta/\partial x^2$; this is

$$\frac{\partial^2\theta}{\partial x^2} \approx$$

Your answer should be $\frac{1}{2} \times 10^{-10}\theta°C\ m^{-2}$ and is obtained from taking the difference between the two values of the gradient and dividing by 100 km , as follows:

$$\frac{\partial^2\theta}{\partial x^2} \approx \frac{\frac{1}{4} \times 10^{-4} - \frac{1}{5} \times 10^{-4}}{10^5}\theta = \frac{1}{2} \times 10^{-10}\theta.$$

Hence $\kappa(\partial^2\theta/\partial x^2) \approx 2\times 10^{-5}\theta$, and this is also the rate of change of temperature with respect to time, $\partial\theta/\partial t$. Crudely, this can be approximated by the difference

$$\frac{\theta(\text{one day later}) - \theta(\text{now})}{24 \text{ hours}},$$

and since 24 hours is 86400 or 9×10^4 s, this gives:

$$\theta(\text{one day later}) = \theta(\text{now})+$$

Your answer should be $\theta(\text{now}) + 1.8°C$. So the temperature one day later is, roughly 12°C. Some readers may have found these last few calculations a new experience; for others their crudity may have been all too obvious. They are an improvement on dimensional analysis, but barely so. However, they do provide an introduction to the use of differences as approximations to gradients (derivatives). The calculation of the second-order gradient $\partial^2\theta/\partial x^2$ is in fact similar to the way in which such terms are computed in numerical finite difference schemes. The latter are, as you may guess, far more sophisticated; so much so in fact that the above crude computations are barely recognisable distant cousins. This has enabled us to reach the stage so that we are able to do a systematic example of numerical prediction. As in the first statistics example, oceanographic relevance has had to be sacrificed for clarity.

Example 8.9 *This example concerns the prediction of the future population of the United States. At a time t, the population of the USA is given by a function P(t). Let t denote the year past 1900, and the function P(t) is assumed to obey the logistic equation:*

$$\frac{dP}{dt} = aP - bP^2.$$

Solution

We will use the forward difference:

$$\frac{dP}{dt} \approx \frac{P(t+h) - P(t)}{h}.$$

The discretised version of the logistic equation using the above forward difference takes the form:

$$P(t+h) \approx$$

This is straightforward to obtain from evaluating the logistic equation at time t with a forward difference to dP/dt. It is:

$$P(t+h) = P(t) + h(aP(t) + bP^2(t)).$$

With data $a = 0.02$, $b = 4 \times 10^{-5}$, $P(0) = 76.1$ and $h = 10$, calculate $P(10)$.

$$P(10) =$$

Your answer should be 89.00 (approximately). Continue with $t = 10$ to find $P(20)$ and so on to complete Table 8.5. The answers you should have obtained are contained within Table 8.6. Note how remarkably good the predictions are. One reason for this is that the parameters a and b are both adjustable and they can be chosen to that the fit between actual and predicted population is minimised. If any reader is around in the year 2020, test the formula then to see how good it is! The writer is not very confident.

Year	t	P(t)(approx)
1920	20	
1930	30	
1940	40	
1950	50	
1960	60	
1970	70	
1980	80	

Table 8.5 *An empty table of population to complete*

Year	t	P(t)(approx)	P(t)(measured)
1900	0	76.10	76.1
1910	10	89.00	92.4
1920	20	103.64	106.5
1930	30	120.97	123.1
1940	40	138.21	132.6
1950	50	158.32	152.3
1960	60	179.96	180.7
1970	70	203.00	204.9
1980	80	227.12	226.5

Table 8.6 *The completed table*

Example 8.10 *This example concerns shallow-water equations and numerical approximations to partial differential equations. These were encountered in Chapters 3 and 6, albeit superficially.*

A simple one-dimensional estuary (or, more strictly, a model that has as its variable a cross-stream average which renders the model one-dimensional) is governed by the equation of motion,

$$\frac{\partial u}{\partial t} + U\frac{\partial u}{\partial x} = -g\frac{\partial \eta}{\partial x} + ku,$$

where u is the speed of the flow, U is a constant background flow, g is acceleration due to gravity, and η is the elevation of the water surface, usually measured from mean sea level (mean low water spring for those with knowledge of tides). The term ku represents in a rather crude way, frictional dissipation. Using only forward differences, put this equation into finite difference form using the notation:

$$\begin{aligned} u_j^s &= u(s\Delta t, j\Delta x), \\ \eta_j^s &= \eta(s\Delta t, j\Delta x), \end{aligned}$$

such that it predicts a new value of u:

$$u_j^{s+1} =$$

Solution
Before giving the answer, there is another relationship which stems from the fact that mass is neither created nor destroyed in the estuary. This takes the rather simple form:

$$\frac{\partial \eta}{\partial t} + h\frac{\partial u}{\partial x} = 0.$$

Here, h is the constant depth of the estuary (the symbol h is universally used for the step length in numerical methods, but we avoid the clash of notation here

by retaining the incremental notation Δx and Δt for the space and time steps
respectively). We also write this mass conservation (called continuity) equation
in finite difference form:

$$\eta_j^{s+1} =$$

There are logical reasons for writing the two equations together. They are:

$$u_j^{s+1} = -U\left(\frac{\Delta t}{\Delta x}\right)(u_{j+1}^s - u_j^s) - g\left(\frac{\Delta t}{\Delta x}\right)(\eta_{j+1}^s - \eta_j^s) + (1 + k\Delta t)u_j^s,$$

$$\eta_j^{s+1} = h\left(\frac{\Delta t}{\Delta x}\right)(u_{j+1}^s - u_j^s) + \eta_j^s.$$

These two equations are solved in tandem starting from the initial conditions.
the main problem with this scheme, which is called *explicit* (because the unknown is on the left of each equation and is explicitly given in terms of known
quantities, this was outlined in Chapter 3. All of this was covered there, it is
repeated here as a reminder). The scheme will be unstable for some values of
Δt and Δx (the time and space step lengths). Another problem with explicit
schemes is their large truncation error, an error due to the poor representation
of the gradients (see Chapter 3). As you can see, even this simplified estuarial
problem involves solving equations that require a computer. Values for Δt, Δx,
U and k are chosen ($g = 9.81\,\mathrm{m\ s^{-2}}$, of course), as well as the start values u_0^0
and η_0^0. This brings us to a third problem. As posed, what happens at the
mouth of the estuary ($x = 0$) dictates what happens upstream. This may not
be entirely appropriate; the message here is *always* ask questions as to the physical nature of what is being modelled and see whether the representation of it is
right. To program this scheme as it stands would certainly be possible, but the
answers would be awry due to the above-mentioned stability, truncation and
design problems. An alternative finite difference formulation using centred differences to avoid instability and higher-order terms to improve accuracy would
be a far more practical modelling approach. All that we get from the above
is an insight, it is hoped a valuable insight, into how finite difference schemes
work.

Here is a simple analytical example that might give insight into geostrophy
and vorticity. It is not practical; very much a learning model.

Example 8.11 *A flow is entirely without North-South gradients. It is initially
such that*

$$\eta = \begin{cases} \eta_0 & x > 0 \\ -\eta_0 & x < 0. \end{cases}$$

*Find the final state given that the flow is geostrophic by using the linearised
potential vorticity equation.*

Solution The problem is that of an infinitely long "bore" along the y-axis which
remains stationary. Physically unrealistic as has been said! Geostrophic flow
gives

$$v = \frac{g}{f}\frac{\partial \eta}{\partial x}, \qquad u = -\frac{g}{f}\frac{\partial \eta}{\partial y} = 0,$$

and the vorticity is:

$$\zeta = \frac{\partial v}{\partial x} - \frac{\partial u}{\partial y} = \frac{g}{f} \frac{\partial^2 \eta}{\partial x^2}.$$

The conservation of potential vorticity (see Chapter 2) is

$$\frac{D}{Dt} \left(\frac{\zeta + f}{h + \eta} \right) = 0$$

which means upon integration that

$$\frac{\zeta + f}{h + \eta} = \begin{cases} \dfrac{f}{h + \eta_0} & x > 0 \\ \dfrac{f}{h - \eta_0} & x < 0. \end{cases}$$

The left hand side of this equation is now expanded in ascending powers of η/h and only linear terms retained. Try this yourself first. You should get the expression

$$\frac{1}{h} \left(\zeta + f - \frac{f}{h} \eta \right).$$

Performing similar expansions on the right thus gives

$$\frac{1}{h} \left(\zeta + f - \frac{f}{h} \eta \right) = \begin{cases} f - \dfrac{f}{h} \eta_0 & x > 0 \\ f + \dfrac{f}{h} \eta_0 & x < 0. \end{cases}$$

Substituting for ζ into this gives a second order differential equation with constant coefficients almost of SHM type which even if used to be unfamiliar, should be no longer after reading the first parts of Chapter 6. Of course, the minus sign means that we will expect exponential rather than oscillatory solutions. The equations to solve are thus

$$\frac{\partial^2 \eta}{\partial x^2} - \frac{f^2}{gh} \eta = \begin{cases} -\dfrac{f^2}{gh} \eta_0 & x > 0 \\ \dfrac{f^2}{gh} \eta_0 & x < 0. \end{cases}$$

These can be solved. Do so if you can. The answers need to be consistent with $\eta \to 0$ as $x \to \pm\infty$ so after some algebra you should get:

$$\eta = \begin{cases} 1 - e^{-x/a} & x > 0 \\ -1 + e^{x/a} & x < 0 \end{cases}$$

where

$$a^2 = \frac{gh}{f^2}$$

and a is therefore the Rossby radius of deformation. Of course $u = 0$ everywhere, but

$$v = \frac{g\eta_0}{af} \begin{cases} e^{-x/a} & x > 0 \\ e^{x/a} & x < 0. \end{cases}$$

Northwards flow is caused by the east to west gradient in the sea surface. You might like to surmise what happens at $x = 0$ the site of the "bore" which has been judiciously omitted from consideration above. Is there a problem?

This next example returns to the Arabian gulf last visited towards the end of Chapter 5. It is also analytical rather than numerical, although its principal use is to validate numerical models; it is too idealised to be of practical use. We abandon the "Example" and "Solution" form for the next two examples as we talk through them calculating as we go. The model is two dimensional with one horizontal co-ordinate and a vertical σ-coordinate. The horizontal density gradient is driving the model, so it resembles an estuary. The basic balance is governed by the equation:

$$\frac{\partial u}{\partial t} - \frac{1}{\rho H^2} \frac{\partial}{\partial \sigma} \left(\nu_z \frac{\partial u}{\partial \sigma} \right) = -g \frac{\partial \eta}{\partial x} + T^{(x)}.$$

In this equation, the σ-coordinate is defined by

$$\sigma = \frac{z + h}{\eta + h} = \frac{z + h}{H},$$

$$T^{(x)} = \frac{g}{\rho} \left[\frac{\partial R}{\partial x} - H(1 - \sigma) \frac{\partial \rho}{\partial x} \right]$$

and

$$R = \int_\sigma^1 [\rho(\sigma) - \rho(\sigma')] d\sigma'.$$

As the density is the driving force behind the model, the continuity and incompressibility equations need careful handling. The gulf occupies the region $0 \le x \le l$ with the open end at $x = l$ and the closed end at $x = 0$. As there is no Coriolis acceleration, there is no underlying geostrophy and hence the flow is confined to be up and down the gulf. Before any calculations can be performed, there are boundary conditions to be assigned and some bulk conditions to be derived. Mass conservation is assured through the general equation

$$\frac{\partial}{\partial x}(\rho u) + \frac{\partial}{\partial y}(\rho v) + \frac{\partial}{\partial z}(\rho w) = 0.$$

However here there is no v and vertical integration is utilised in order to eliminate z. With steady flow therefore the continuity equation reduces to

$$\int_{-h}^\eta \rho u \, dz = H \int_0^1 \rho u \, d\sigma = 0.$$

See if you can derive this from integrating the previous equation. The x derivative of the above integral is zero from the integral of the continuity equation.

The above equation arises because there is no flow through the closed end. It is assumed that the eddy viscosity ν_z is constant. At the sea bed, it is assumed that

$$\nu_z \frac{\partial u}{\partial \sigma} = \bar{\rho} h C_D u^{(b)}.$$

This is the standard linear bottom friction which takes the form $\bar{\rho} C_D u^{(b)}$ where

$$\bar{\rho} = \int_0^1 \rho \, d\sigma$$

is the vertically averaged density. The left hand side $\nu_z \frac{\partial u}{\partial \sigma}$ is the bottom stress. There is no wind stress, so what do you think is an appropriate surface condition? It is

$$\frac{\partial u}{\partial \sigma} = 0.$$

Insert our assumptions into the single momentum equation, and see if you can derive:

$$\frac{\partial}{\partial \sigma}\left(\nu_z \frac{\partial u}{\partial \sigma}\right) = g\rho h^2 \frac{\partial \eta}{\partial x} + g h^3 \frac{\partial}{\partial x}\int_0^1 \rho(\sigma') d\sigma' - g h^2 \frac{\partial h}{\partial x}\int_0^1 [\rho(\sigma) - \rho(\sigma')] d\sigma'.$$

Insert the specific density profile $\rho = \rho_0(1 - \gamma x)$ and assume $h = $ constant and find the equation for u. You should get

$$\frac{\partial}{\partial \sigma}\left(\nu_z \frac{\partial u}{\partial \sigma}\right) = g\rho h^2 \frac{\partial \eta}{\partial x} + \gamma g h^3 (1 - \sigma).$$

By using the boundary conditions, see if you can integrate the above equation twice with respect to σ to obtain an explicit formula for u. The result is

$$u = \frac{U}{1 + 4k}[1 + 12k\sigma - 6(1 + 5k)\sigma^2 + 4(1 + 4k)\sigma^3],$$

and

$$\frac{\partial \eta}{\partial x} = \frac{1}{2}\gamma h \frac{1 + 3k}{1 + 4k},$$

where

$$k = \frac{C_D h \rho_0}{12\nu_z}, \qquad U = \frac{\gamma g h^3 \rho_0}{24\nu_z}.$$

For the values $\rho_0 = 1035 kg.m^{-3}$ and $\gamma = 1.47 \times 10^{-8} m^{-1}$ and using the values:

$$g = 9.81, \quad C_D = 0.002, \quad \nu_z = 0.065, \quad l = 800 km.$$

calculate the actual density at each end of the channel. The answers are $1035 kg.m^{-3}$ at the open end, and $1025 kg.m^{-3}$ at the closed end. The flow is not dependent upon x, but calculate its magnitude. It is about $20 mm.s^{-1}$. Of course this model is extremely simple, the use of finite difference methods becomes necessary for less idealised density distributions.

Let us now move unashamedly into the field of numerical modelling. It was mentioned in Chapter 3 that the sophisticated models outlined there require validating, in particular models of fronts are controversial because there is always some numerical diffusion in a finite difference model. As fronts are sharp gradients in density (and usually current too) trying to model these is a good test for any scheme. The model has to keep the integrity of the front as it moves through the sea, and a simple test of a scheme therefore is to solve the pure advection equation

$$\frac{\partial C}{\partial t} + u\frac{\partial C}{\partial x} = 0$$

written in one dimension for convenience. This is of course, using differentiation following the motion (Lagrangian picture)

$$\frac{DC}{Dt} = 0$$

which integrates to $C =$ constant following the motion. Attempting to solve the equation

$$\frac{\partial C}{\partial t} + u\frac{\partial C}{\partial x} = 0$$

numerically therefore will give a good indication of how the scheme preserves the shape of the initial distribution of concentrate. The first scheme we test is to use a simple upstream gradient to approximate the advection:

$$\frac{\partial u}{\partial x} = \frac{C_j - C_{j-1}}{\Delta x}.$$

This assumes that the concentrate is advected with speed C. With the numerical values shown in the caption, the advection of a Gaussian patch is shown in Figure 8.10 where all calculations are performed with a Courant number of 0.2. (The Courant number, defined in Chapter 3 is the dimensionless ratio of actual current speed to numerical speed $\Delta x/\Delta t$). Here it is $C\Delta t/\Delta x$. It is seen that (a) contains to much diffusion, (b) is better but (c) is best of all. There is always some oscillation with respect to the background. Note that when conducting experiments such as this, there has to be a finite background value of C, usually $C = 1$ otherwise any oscillation will result in negative values. In order to preserve sharp fronts we try the strongly non-linear form:

$$\frac{\partial u}{\partial x} = \frac{C_j^4 - C_{j-1}^4}{\Delta x}.$$

Here we have taken $u = C^4$ and plotted the results in Figure 8.11 where $t = 0.1$ and the time steps were for (a) 0.0005 and for (b) 0.0001. The front is quite well preserved, but centred differencing works a little better for this strongly non-linear problem. The author in indebted to Johan de Kok (Rijkwaterstaat, The Netherlands) for access to these simple test examples for numerical advection schemes. Those of James, mentioned in Chapter 3 were developed from such simple tests although they are now of course very much more complicated.

Here is an idealised, very idealised ecosystem example.

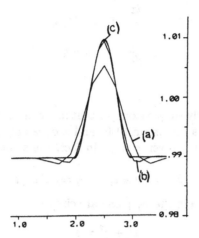

Figure 8.10: First numerical solutions to the advection equation. (a) $\Delta x = 0.2016$, (b) $\Delta x = 0.1028$, (c) $\Delta x = 0.01$

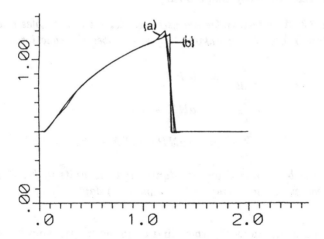

Figure 8.11: Second numerical solutions to the advection equation. (a) $\Delta x = 0.104$, (b) $\Delta x = 0.02$

Example 8.12 *Investigate the predator prey system*

$$\frac{dx}{dt} = -x - y$$
$$\frac{dy}{dt} = x - y$$

Solution

This is a particularly simple example that in fact is easily solved exactly. The centre of the motion is the origin $(0,0)$ and x does not grow wherever $x = -y$ and y does not grow wherever $x = y$. In fact, the solution for x and y is the curve

$$x = e^{-t}\cos t, \qquad y = e^{-t}\sin t$$

so that the trajectories in the xy-plane are circles

$$x^2 + y^2 = e^{-2t}$$

which have radius e^{-t}. This means that as time progresses the circles get smaller and smaller eventually being so close to the origin as to be indistinguishable from it. The origin is thus the centre and x and y eventually die out.

Not very much in the way of practical deductions can be made from such idealised models, but general methods (plotting x against y, seeing how each varies with time etc.) are always useful to learn.

Our next example examines a simple is a little more practical and examines a simple but realistic ecosystem model.

Example 8.13 *A three-variable ecosystem model has as variables nutrient (N), phytoplankton (P), and zooplankton (Z), which obey the equations:*

$$\frac{dZ}{dt} = b_2 PZ - dZ,$$
$$\frac{dP}{dt} = aNZ - bPZ,$$
$$\frac{dN}{dt} = -aNP + b_1 PZ + dZ,$$

where $b = b_1 + b_2$, a and d are constants, and the units of N, Z and P are milligrams per cubic metre (mg m^{-3}). Time is in days.

Solution

By adding these equations together, find a simple relationship between Z, N and P:

You will find that the right-hand side adds to zero, since if the gradient of a quantity is zero the quantity does not change, which means that the quantity is constant. Hence $Z + P + N = $ constant. Take this constant as 5.0 (see the paper by Klein and Steele (1985) from which this example is derived). One possible state is $N = 5$, $Z = P = 0$, which is called a steady state solution since it does

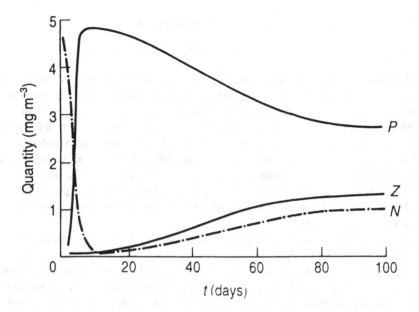

Figure 8.12: The variation of P, Z and N with time. From Klein and Steele (1985). Reproduced with permission of the authors.

not change with time. It is also an entirely feasible solution, if a very boring one! Is it the only one? To answer this question, put all the right-hand-side rates of change equal to zero, do not allow any of the variables Z, P or N to be zero, and solve the three simultaneous linear equations:

$$Z = \qquad P = \qquad N =$$

The solution to these equations are:

$$P = d/b_2,$$
$$Z = \left(5 - \frac{d}{b_2}\right) / \left(1 + \frac{b}{a}\right),$$
$$N = \left(5 - \frac{d}{b_2}\right) / \left(1 + \frac{b}{a}\right).$$

We are now in a position to experiment with this model a little, although what we can do here is limited by what can be done without running to a computer for help with the solving of the equations. Let us assume that, at time $t = 0$, $N = 5$, $Z = 0$ and $P = 0$. Also assume that, after a long time, the variables Z, P and N reach the above steady state values. First, calculate these given the following values of the constants that appear in the model: $a = 0.2$, $b = 0.15$, $b_2 = 0.03$, and $d = 0.08$.

$$Z = \qquad P = \qquad N =$$

The arithmetic should reveal the following values:

$$Z = 1.33, \qquad P = 2.67, \qquad N = 1.00.$$

However, we are unable to deduce precisely *how* Z, P and N reach these values; this can only be done by solving the equations using for example a marching method (see example 8.8). The actual shapes of the variations of the three variables are shown in Figure 8.12. P is low at the start but rapidly climbs to a value close to its maximum (5.0). At this time, both nutrient (N) and zooplankton (Z) values are close to zero. After this time, all variables tend to their predicted eventual steady state values $P = 2.67$, $Z = 1.33$ and $N = 1.0$.

As mentioned earlier, this example has been adapted from the paper by Klein and Steele (1985), in which the role of introducing diffusion into the model is considered in some detail.

Finally, from one extreme to the other, here is an example that has very much an engineering flavour. It concerns the prediction of extreme waves. It is a little different from what has gone before and could have slotted into the end of the last chapter. As yet there is no purely theoretical method for predicting how waves evolve from the wind. Instead reliance is upon empirically derived formula. In Chapter 7 the PM (Pierson Moskowitz) and JONSWAP spectra were introduced. The significant wave height (mean of the highest one third) was defined and given the symbol H_s. Let T_s be the period associated with this wave. The following empirically derived formulae relate these to the speed of the wind at 10m above the sea

$$H_s \quad \sim \quad 0.025 U_{10}^2$$
$$T_s \quad \sim \quad 0.79 U_{10}$$

for the PM spectrum, and

$$H_s \quad \sim \quad 5.1 \times 10^{-4} U_{10} F^{0.5}$$
$$T_s \quad \sim \quad 0.059 \{U_{10} F\}^{0.33}$$

for the JONSWAP spectrum. In the latter, F is the fetch as it will be remembered that this spectrum is for use with non-fully developed seas. Although the above relationships are empirical, they have been arrived at by engineers and are the best that can be done under present knowledge. Engineers have codes of practice that govern how they can design and predict from a legal standpoint. The above formulae are part of this code. Calculate the significant wave height and period for a wind of $20ms^{-1}$ at a height of 10m above the sea if the sea is fully developed.

$$H_s \quad \sim$$
$$T_s \quad \sim \qquad .$$

The PM formulae should have been used with answers $H_s = 10m$, and $T_s = 15.8s$. Suppose now that the fetch is limited to the width of the North Sea which is about 450km. Find the new values of H_s and T_s.

$$H_s \quad \sim$$
$$T_s \quad \sim \quad .$$

This time, the JONSWAP spectrum is appropriate and these formulae give $H_s = 6.8m$ and $T_s = 12.3s$. Note that these values are significantly less than the fetch unlimited values. Most may think that the North Sea must be wide enough to "wind up" the sea surface; it isn't. The two spectra only give the same values if a fetch of close to 1000km is assumed. Of course, this is very simplified. How much fetch is required depends crucially on the value of U_{10} and we have said nothing about limiting the duration of the wind.

Figure 8.13 shows a straight line arising from many measurements and forms part of BS6349. We will try to use the straight line to predict 1 year, 50 year and 100 year significant wave heights. This graph is actually based on the Weibull distribution. The factor p denotes what is called the "exceedence probability". This probability is linked to the return period through the formula

$$p = \frac{1}{T}.$$

Engineers have to be able to predict long term waves with incomplete data sets. The methods used are based on sound enough statistical techniques, but the margins of error, hardly ever referred to are high. Nevertheless here is what is done. What wave record available are analysed and all waves that exceed a threshold value are subtracted out. For this example, suppose that in a year long record there are 39 "storms", that is 39 times that the threshold value has been exceeded. The exceedence probability is then given by

$$p_1 = \frac{1}{39 + 1} = \frac{1}{40}$$

for this data. It is then a simple multiplication factor. The exceedence probability for a 50 year wave is then

$$p_{50} = \frac{1}{50} \frac{1}{40} = \frac{1}{2000}$$

and the exceedence probability for the 100 year wave is half this

$$p_{100} = \frac{1}{4000}.$$

The probability theory behind these formulae are outside the scope of this text but can be found in statistics texts that include the practical use of probability. Assuming the values of p given above, now calculate the wave heights.

The answers are:

$$\begin{aligned}
1 \text{ year wave} &= 3.9m \\
50 \text{ year wave} &= 7.3m \\
100 \text{ year wave} &= 7.9m
\end{aligned}$$

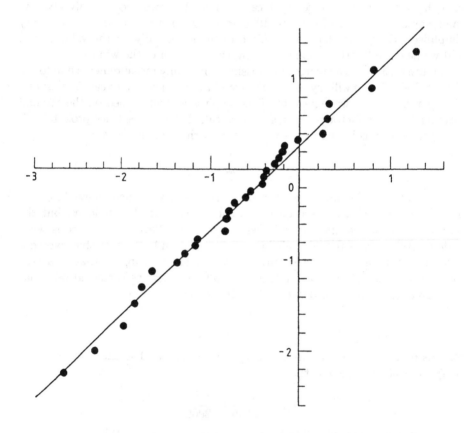

Figure 8.13: A wave height extrapolation plot: The line has equation
$ln\{ln(1/p)\} = 0.945 ln\{H_s - 0.95\} + 0.278$

Now, if global warming provides a stormier environment, there is a good case for seeing what happens to these predictions when there are more storms. Suppose the rate increases from 39 to 49 storms each year, calculate the new wave heights for 1 year, 50 year and 100 year return periods.

The answers are:

$$1 \text{ year wave } = 4.1m$$
$$50 \text{ year wave } = 7.5m$$
$$100 \text{ year wave } = 8.1m$$

which, given that the energy is proportional to the square of the wave height could give designers of offshore platforms and the like headaches! Even more bizarre is the attempted forecast of highest waves. If extreme value statistics are used to predict 3 hour maximum waves, the Gumbel distribution comes up with waves of height $H_{max} \sim 30m$ for the northern part of the North Sea. Of course this is based on records that are too short (3 years) and so the findings are suspect. No waves of this height have been seen there and no structure yet built there could survive such a wave. Does it need to?

It is hoped that the examples in this chapter have, at least in some small way, helped you to understand what is happening when the process of mathematical modelling takes place. It is recognised that many of the examples are so idealised that they are of little use to those who model or wish to model real marine systems. Such real-life modelling is the logical next step for many of you, but to be successful in that endeavour may well require more technical knowledge than can be found in this text. For those who only wish to use models, there is enough here to give you some insight into what is behind a number of different types of marine model and to question its operation critically.

8.4   Exercises

Here are some problems for you to try. Some are idealised, others less so. The answers are given at the end of the book.

1. Find the mode, median and (arithmetic) mean of the following sets of numbers:

   (a) $3, 5, 7, 5, 1, 9, 5, 4, 8, 2, 3, 7, 4, 9, 5, 1, 6, 2, 3, 1.$
   (b) $23, 4, 42, 76, 59, 23, 11, 51, 87, 99, 46, 62, 69, 36, 59, 14, 1, 15, 82, 94.$
   (c)

Height(inches)	Mark (X)	Frequency (f)	fX
60-62	61	5	305
63-65	64	18	1152
66-68	67	42	2814
69-71	70	27	1890
72-74	73	8	584

Table 8.7

(d)

Weight(mg)	Mark (X)	Frequency (f)	fX
0-10	5	4	20
11-20	15.5	11	170.5
21-30	25.5	31	790.5
31-40	35.5	53	1881.5
41-50	45.5	41	1865.5
51-60	55.5	48	2664
61-70	65.5	29	1899.5
71-80	75.5	18	1359
81-90	85.5	8	684

Table 8.8

2. Calculate the standard deviation for each of the four sets of data given in Exercise 1.

3. Table 8.9 gives actual and expected catches of fish over a ten day period

	Species A	Species B	Species C	Species D	Species E
Expected	370	81	46	159	5
Actual	380	79	31	146	6

Table 8.9

Using a χ^2 test with $\alpha = 0.01$, determine whether the actual catch deviates significantly from expectation. If $\alpha = 0.75$, which corresponds to a 25% level of significance is the expectation significant now? A new set of data is given in Table 8.10.

	Species A	Species B	Species C	Species D	Species E
Expected	370	81	46	159	5
Actual	369	31	49	148	12

Table 8.10

What are the conclusions now? Comment on the interpretation of the statistics according to the χ^2 test as opposed to the interpretation you get from "eyeballing" the data.

4. The data given in Table 8.11 show the discharges of nitrogen (N) and phosphorus from a river. The total discharge (Q) is also given. Calculate the correlation coefficients as well as the two regression lines N on Q and P on Q.

Discharge	1981	1982	1983	1984	1985	1986	1987
N(t/yr)	14000	8600	6400	13800	12700	11200	10900
P(t/Yr)	210	190	140	190	200	160	170

Discharge (contd)	1988	1989	1990	1991	1992	1993
N(t/yr)	16100	7900	12000	13800	6000	15300
P(t/Yr)	220	120	190	120	90	220

Table 8.11

5. The distribution of cadmium along the centre line of a polluted river from the estuary upstream is presented in two ways. As a table (Table 8.12) and as a graph, Figure 8.14. Calculate the mean, mode and median from the table and indicate how you would estimate each of these from Figure

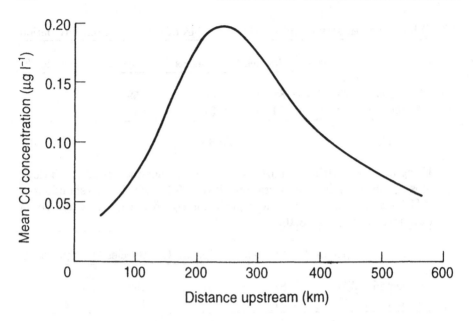

Figure 8.14: Variation of Cd with distance upstream

8.14 *without* the use of Table 8.12. Which of these measures of central tendency gives the clearest indication of the position of the source of the cadmium?

Distance upstream (km)	Mean Cd concentration (μgl^{-1})
0–100	0.042
101–200	0.121
201–300	0.195
301–400	0.136
401–500	0.082

Table 8.12

How would you expect the variance (as defined in example 8.1) of the distribution of cadmium to change with time, give that no further cadmium is being discharged into the river?

6. Consider the following facts about the Mediterranean Sea: the salinity of the in-flowing water is 36.2 parts per thousand, the salinity of the out-flowing water is 38.4 parts per thousand, its surface area is $2.5 \times 10^{12} m^2$ and the amount of freshwater evaporated per year is 1 m. If Q_1 is the quantity of water entering the Mediterranean and Q_2 the quantity going out, by writing down equations for overall salt balance and water balance,

calculate Q_1 and Q_2. [This is adapted from a problem in George Mellor's excellent little book, Mellor (1996)]

7. An ocean current is 200km wide with a speed of $1ms^{-1}$ at a latitude of 45°N. If the lateral friction has magnitude $2 \times 10^6 m^2 s^{-1}$ use dimensional methods to outline the likely dynamic balance. How does this change if the current narrows to 50km., the current speed increases to $4ms^{-1}$ and the friction decreases to $2 \times 10^5 m^2 s^{-1}$?

   Are any of the models outlined in Chapter 2 appropriate to use in either of these cases? If so, which ones and if not can you develop a possible model from the principal balance.

8. A tidal current has magnitude $5ms^{-1}$ in an estuary whose width narrows from 10 to 5 km. The tidal range is 2m., the period is 12hrs., and the latitude is 56°N. Friction is known to be important. Assuming the quadratic law $\rho C_D U^2$ where ρ is the density, u is the current speed and C_D is a constant, determine an appropriate magnitude for C_D.

9. Consider once more the Kelvin wave (see section 6.5.1). The straight coast is the x axis with the sea occupying the region $y \geq 0$. The longshore flow is in geostrophic balance:

$$u = -\frac{g}{f}\frac{\partial \eta}{\partial y}$$

   but this time there is linear friction and the momentum equation is

$$\frac{\partial u}{\partial t} = -g\frac{\partial \eta}{\partial x} + ku$$

   where k is a friction coefficient, and the continuity equation (mass conservation) remains

$$\frac{\partial \eta}{\partial t} + h\frac{\partial u}{\partial x} = 0.$$

   Determine the solution by assuming solutions proportional to $\exp\{i\omega t + i\alpha x\}$, find the constant α then expand in the small parameter k/ω and hence show that the length scale for the decay of the waves is

$$\sqrt{\frac{gh}{k\omega}}.$$

10. The semi diurnal lunar tide M_2 has a period 12.42 hours, and the semi diurnal solar tide S_2 has a period 12.00 hours. By assuming both are sinusoidal waves of the same amplitude, write down the combination and deduce the period of "spring" tides. What does this correspond to physically?

11. Suppose that there is a continental shelf along a curved coastline. Shelf waves are being modelled. The curve is considered as an arc of a circle, and polar co-ordinates are employed. The equation obeyed by the streamfunction ψ, related to the longshore current v through

$$vH = \frac{d\psi}{dr}$$

is

$$\frac{1}{r}\frac{d}{dr}\left(r\frac{d\psi}{dr}\right) - \frac{1}{H}\frac{dH}{dr}\frac{d\psi}{dr} = H^2 G(\psi) - Hf.$$

Take $G(\psi) = G_0 e^{-\mu\psi}$, $H = H_0 e^{\lambda(r-r_0)}$ where H_0, G_0, λ and μ are constants. Determine the dimensions of G_0 and μ, identify typical length and time scales. Use dimensional analysis to identify two dimensionless groups. The first containing H_0, G_0, λ, μ with the term $H_0^2\mu$. The second containing H_0, G_0, λ, μ and f with the term $\mu f H_0$. Assuming that each of these are of order unity determine the magnitudes of H_0 and G_0 given the following magnitudes: $\mu = 4 \times 10^{-8} m^{-3} s$, $\lambda = 10^{-5} m^{-1}$ and $f = 5 \times 10^{-5} s^{-1}$. Write down the non-dimensional version of the differential equation for ψ.

Given that the model is for the longshore flow due to continental shelf waves, describe physically what is probably happening at the locations where

$$\frac{d\psi}{dr} = 0.$$

[This is based on the work of Roger Hughes, see Hughes (1989) and reference cited there.]

12. In a small square inlet occupying the region $-a \le x \le a$, $0 \le y \le 2a$ the surface elevation η and velocity (u,v) satisfy

$$\frac{\partial u}{\partial t} = -g\frac{\partial \eta}{\partial x},$$
$$\frac{\partial v}{\partial t} = -g\frac{\partial \eta}{\partial y},$$
$$\frac{\partial \eta}{\partial t} + h\left(\frac{\partial u}{\partial x} + \frac{\partial v}{\partial y}\right) = 0.$$

Looking for solutions relevant to tidal oscillations (rotational effects due to Coriolis acceleration are ignored) let

$$u = U(x,y)\cos\omega t, \quad v = V(x,y)\cos\omega t, \quad \eta = A(x,y)\sin\omega t$$

and hence deduce an equation satisfied by $A(x,y)$. Suggest possible solutions if all sides are coasts except that at $y = 0$ where there is an imposed tidal oscillation $\eta = \eta_0 \cos\omega t$.

13. The differential equation

$$\frac{dy}{dt} = 0.1y + 0.02y^2 + t$$

arises from a population model of Antarctic krill. y is the (scaled) biomass and t is of course time. By using a forward difference approximation for the left-hand side with a step size of 0.1 predict the value of y at time $t = 0.5$ given that $y = 1$ when $t = 0$. Some of you can use sophisticated methods, most will probably utilise Euler's method as done in Example 8.8.

Using a very accurate method, y is predicted to be 1.19. Compare your result with this and discuss the reasons for any difference.

14. The differential equation

$$\frac{dy}{dt} = \frac{0.1x}{5t + x}$$

arises from a model that has been built to try and predict the growth of an algal patch which is subject to self shadowing (x is the patch size and t is time). Solve this equation as in the last exercise using a step size of 10.0 and predict the size of the patch at time $t = 50$ given that $x = 100$ when $t = 0$.

Explain any differences between your answer and the more accurate answer $102.5m$?

15. The diffusion of salt $S(x, t)$ in a one dimensional estuarial model obeys the advection-diffusion equation

$$\frac{\partial S}{\partial t} + U\frac{\partial S}{\partial x} = \kappa\frac{\partial^2 S}{\partial x^2},$$

where U is a constant flow, κ is a constant diffusion coefficient, x is measured upstream from the estuary and t is time. Adopting the notation $S_{i,j} = S(i\Delta t, j\Delta x)$ write this equation in finite difference form using the centred difference in space for the second derivative and a forward difference in time.

Outline the difficulties with his method, and discuss possible better alternatives. How are boundary values incorporated remembering that the model is trying to predict the salinity along the centre line of the estuary?

16. The simplest two-dimensional partial differential equation is called Laplace's equation and can be written

$$\frac{\partial^2 \phi}{\partial x^2} + \frac{\partial^2 \phi}{\partial y^2} = 0.$$

Taking equal steps in the x and y directions, show that the finite difference form states that at any interior grid point the value of the variable ϕ is the arithmetic mean of the values of ϕ at the surrounding grid points. Suppose that this equation is to be solved in an area consisting of 10,000 grid points, what extra information is needed before we can solve the problem? If the finite difference form of Laplace's equation for each of the 10,000 points is written in matrix form

$$Ax = b,$$

what would be the general form of the matrix A?

17. Here is longer "case study" type of problem concerning the modelling of pollution in an inlet. The basic equation is the advection-diffusion equation written in flux form:

$$\frac{\partial C}{\partial t} + \frac{\partial}{\partial x}(uC) + \frac{\partial}{\partial y}(vC) = \kappa \left(\frac{\partial^2 C}{\partial x^2} + \frac{\partial^2 C}{\partial y^2} \right) - \lambda C.$$

Write down finite difference forms for the two second order partial derivatives on the right hand side. Suggest upwind difference forms for the two advection terms

$$\frac{\partial}{\partial x}(uC) \quad \text{and} \quad \frac{\partial}{\partial y}(vC).$$

Consider a perfectly square inlet as shown in Figure 8.15. There are 16 interior points. The square is $10km \times 10km$ with $\Delta x = \Delta y = 2km$. The parameter values are $\lambda = 10^{-3}$ and $\kappa = 4000$.

(a) Suppose the current is everywhere zero, and that initially all values of $C = 1$ apart from the following four:

$$C_{4,4} = C_{6,4} = C_{4,6} = C_{6,6} = 2.$$

Use a forward difference in time with step $\Delta t = 10$mins. to calculate $C_{i,j}^n$ at subsequent times. (Four time steps is reasonable.)

(b) Use the same values of C as in part (a) however specify the following constant velocities:

$$v_{2,4} = v_{2,6} = v_{2,8} = 1, \; v_{8,4} = v_{8,6} = v_{8,8} = -1$$

and

$$u_{4,2} = u_{6,2} = u_{8,2} = -1, \; u_{2,4} = u_{2,6} = u_{2,8} = 1.$$

Once more use forward differences in time with the same scheme as in (a) to calculate four values of $C_{i,j}^n$. Compare the two. What is the Courant number here; are the schemes stable? Are they reasonably free of truncation error?

You might like to experiment with this problem. If you can program it, do so and refine the scheme.

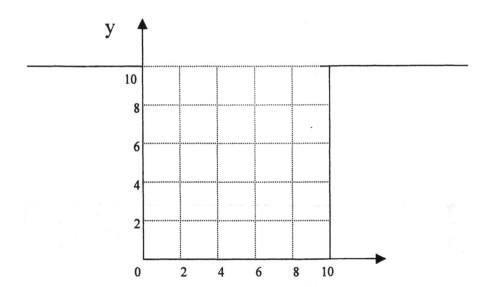

Figure 8.15: The idealised square inlet

18. The graph shown as Figure 8.16 has been produced using different (Norwegian) data to the extreme wave extrapolation plot of Figure 8.13. What does this predict for the 1 year, 50 year and 100 year return period wave (significant wave height, $H_s = H_{1/3}$)? Are the differences between the two predictions significant and/or explicable? (Note that this graph also gives predictions of maximum wave height H_{max} that go above $30m$. The famous "100 foot waves" seen from time to time by those who sail in the antarctic around Cape Horn.)

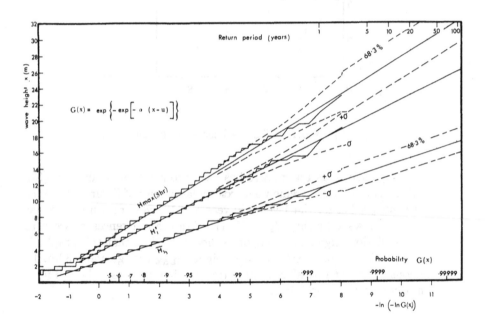

Figure 8.16: A chart, based on three years of wave records and the Gumbel distribution for wave prediction

Answers to Exercises

1. (a) 4.5, (b)47.65, (c)67.65, (d)46.64.

2. (a) 2.565, (b)31.00, (c)2.935, (d)18.096.

3. $\chi^2 = 2.092$. Not significant at the 99% level; significant at the 25% level. New table, $\chi^2 = 41.62$. Significant at both levels due to outliers in the data.

4. Correlation Coefficient $= 0.7281$ Regression Line is $y = 65.55 + 0.0092x$. Correlation Coefficient $= 0.708$ Regression Line is $y = 6.537 + 0.0065x$.

5. The equations satisfied by Q_1 and Q_2 are:

$$0.0362Q_1 = 0.0384Q_2, \quad \text{and} \quad Q_1 - Q_2 = 2.5 \times 10^{12},$$

with solution $Q_1 = 4.11 \times 10^{13} m^3 s^{-1}$, $Q_2 = 4.36 \times 10^{13} m^3 s^{-1}$.

6. Mean $= 267,^*$ mode $= 250$, median $= 264$. The variance would increase, theoretically proportional to time. (*This is the mean *of the distribution*, not the mean concentration which is 0.1152, the average concentration over the estuary.)

7. Horizontal friction versus Coriolis acceleration (large horizontal Ekman number, small Rossby number). New figures, advection versus Coriolis acceleration (small horizontal Ekman number, large Rossby number).

8. 10^9 in S.I. units.

9.

$$\alpha^2 = \frac{\omega}{gh}(\omega + ik)$$

$$\eta = \eta_0 \exp\left\{-\frac{fy}{\sqrt{gh}} - \frac{kx}{\sqrt{gh}}\right\} \cos\left\{\frac{\omega x}{\sqrt{gh}} + \frac{kfy}{\omega\sqrt{gh}} + \omega t\right\}$$

$$u = -\frac{g}{f}\frac{\partial \eta}{\partial y}.$$

10.

$$\cos\omega_1 t + \cos\omega_2 t = 2\cos\frac{\omega_1 t + \omega_2 t}{2}\cos\frac{\omega_1 t - \omega_2 t}{2},$$

Longer period is 29.57 days corresponding to the frequency $(\omega_1 t - \omega_2 t)/2$. Physically, these peaks occur at new and full moon.

11. Length scale: $\sim r_0$, $G_0 \sim L^{-1}T^{-1}$, $\mu \sim L^{-3}T$. Dimensionless groups: $\mu H_0^2 G_0 \lambda^{-2}$, $\mu f H_0 \lambda^{-2}$. $H_0 = 50m$, $G_0 = 10^{-6}m^{-1}s^{-1}$.

$$\frac{1}{r}\frac{d}{dr}\left(r\frac{d\psi}{dr}\right) - \frac{d\psi}{dr} = \mu H_0^2 G_0 \lambda^{-2}\exp\{-\psi + 2(r-r_0)\} - \mu f H_0 \lambda^{-2}\exp\{r-r_0\},$$

where r and ψ are now dimensionless.

12. Possible solutions:

$$\eta = \eta_0 \sin\frac{\pi x}{a}\cos\frac{\pi y}{4a}\cos\omega t$$

where

$$\omega^2 = gh\left(\frac{\pi^2}{a^2} + \frac{\pi^2}{(4a)^2}\right) = \frac{17}{16}\pi^2 gh.$$

Use sin where there are two solid boundaries and cos where there is one solid and one water boundary. The above is the fundamental mode, other higher modes are also of course solutions.

13. 1.16. Truncation error.

14. 102.92. Reasonably satisfied.

15.

$$S_{i,j+1} = S_{i,j} - \frac{U\Delta t}{2\Delta x}(S_{i,j+1} - S_{i,j-1}) + \frac{\kappa\Delta t}{(2\Delta x)^2}(S_{i,j+1} - 2S_{i,j} + S_{i,j-1})$$

16. Boundary conditions.

$$A = \begin{pmatrix} -4 & 1 & 0 & 0 & \cdots \\ 1 & -4 & 1 & 0 & \cdots \\ 0 & 1 & -4 & 1 & \cdots \\ 0 & 0 & 1 & -4 & \cdots \\ \vdots & \vdots & \vdots & \vdots & \end{pmatrix} \quad \text{a banded matrix.}$$

17. Courant number not well defined, but $= 0.3$ for unit wave speed. When there is current, the Courant number may increase but not significantly. There is some truncation error.

18.

$$H_s \text{ (one year)} = 11.5m$$
$$H_s \text{ (fifty year)} = 15m$$
$$H_s \text{ (one hundred year)} = 17m.$$

These results are very different from the predictions of Figure 8.13. Reasons include: different (more violent?) sea area; lots of uncertainty, Figure 8.16 gives error bounds but the value of the distribution $(G(x))$ is very sensitive to the ill derived parameter α.

Appendix A

Percentile values (χ_p^2) for the χ^2 distribution, with v degrees of freedom

					Levels of significance								
v	$\chi^2_{0.995}$	$\chi^2_{0.99}$	$\chi^2_{0.975}$	$\chi^2_{0.95}$	$\chi^2_{0.90}$	$\chi^2_{0.75}$	$\chi^2_{0.50}$	$\chi^2_{0.25}$	$\chi^2_{0.10}$	$\chi^2_{0.05}$	$\chi^2_{0.025}$	$\chi^2_{0.01}$	$\chi^2_{0.005}$
1	7.88	6.63	5.02	3.84	2.71	1.32	0.455	0.102	0.0158	0.0039	0.0010	0.0002	0.0000
2	10.6	9.21	7.38	5.99	4.61	2.77	1.39	0.575	0.211	0.103	0.0506	0.0201	0.0100
3	12.8	11.3	9.35	7.81	6.25	4.11	2.37	1.21	0.584	0.352	0.216	0.115	0.072
4	14.9	13.3	11.1	9.49	7.78	5.39	3.36	1.92	1.06	0.711	0.484	0.297	0.207
5	16.7	15.1	12.8	11.1	9.24	6.63	4.35	2.67	1.61	1.15	0.831	0.554	0.412
6	18.5	16.8	14.4	12.6	10.6	7.84	5.35	3.45	2.20	1.64	1.24	0.872	0.676
7	20.3	18.5	16.0	14.1	12.0	9.04	6.35	4.25	2.83	2.17	1.69	1.24	0.989
8	22.0	20.1	17.5	15.5	13.4	10.2	7.34	5.07	3.49	2.73	2.18	1.65	1.34
9	23.6	21.7	19.0	16.9	14.7	11.4	8.34	5.90	4.17	3.33	2.70	2.90	1.73
10	25.2	23.2	20.5	18.3	16.0	12.5	9.34	6.74	4.87	3.94	3.25	2.56	2.16
11	26.8	24.7	21.9	19.7	17.3	13.7	10.3	7.58	5.58	4.57	3.82	3.05	2.60
12	28.3	26.2	23.3	21.0	18.5	14.8	11.3	8.44	6.30	5.23	4.40	3.57	3.07
13	29.8	27.7	24.7	22.4	19.8	16.0	12.3	9.30	7.04	5.89	5.01	4.11	3.57
14	31.3	29.1	26.1	23.7	21.1	17.1	13.3	10.2	7.79	6.57	5.63	4.66	4.07
15	32.8	30.6	27.5	25.0	22.3	18.2	14.3	11.0	8.55	7.26	6.26	5.23	4.60
16	34.3	32.0	28.8	26.3	23.5	19.4	15.3	11.9	9.31	7.96	6.91	5.81	5.14
17	35.7	33.4	30.2	27.6	24.8	20.5	16.3	12.8	10.1	8.67	7.56	6.41	5.70
18	37.2	34.8	31.5	28.9	26.0	21.6	17.3	13.7	10.9	9.39	8.23	7.01	6.26
19	38.6	36.2	32.9	30.1	27.2	22.7	18.3	14.6	11.7	10.1	8.91	7.63	6.84
20	40.0	37.6	34.2	31.4	28.4	23.8	19.3	15.5	12.4	10.9	9.59	8.26	7.43
21	41.4	38.9	35.5	32.7	29.6	24.9	20.3	16.3	13.2	11.6	10.3	8.90	8.03
22	42.8	40.3	36.8	33.9	30.8	26.0	21.3	17.2	14.0	12.3	11.0	9.54	8.64
23	44.2	41.6	38.1	35.2	32.0	27.1	22.3	18.1	14.8	13.1	11.7	10.2	9.26
24	!45.6	43.0	39.4	36.4	33.2	28.2	23.3	19.0	15.7	13.8	12.4	10.9	9.89
25	46.9	44.3	40.6	37.7	34.4	29.3	24.3	19.9	16.5	14.6	13.1	11.5	10.5
26	48.3	45.6	41.9	38.9	35.6	30.4	25.3	20.8	17.3	15.4	13.8	12.2	11.2
27	49.6	47.0	43.2	40.1	36.7	31.5	26.3	21.7	18.1	16.2	14.6	12.9	11.8
28	51.0	48.3	44.5	41.3	37.9	32.6	27.3	22.7	18.9	16.9	15.3	13.6	12.5
29	52.3	49.6	45.7	42.6	39.1	33.7	28.3	23.6	19.8	17.7	16.0	14.3	13.1
30	53.7	50.9	47.0	43.8	40.3	34.8	29.3	24.5	20.6	18.5	16.8	15.0	13.8
40	66.8	63.7	59.3	55.8	51.8	45.6	39.3	33.7	29.1	26.5	24.4	22.2	20.7
50	79.5	76.2	71.4	67.5	63.2	56.3	49.3	42.9	37.7	34.8	32.4	29.7	28.0
60	92.0	88.4	83.3	79.1	74.4	67.0	59.3	52.3	46.5	43.2	40.5	37.5	35.5
70	104.2	100.4	95.0	90.5	85.5	77.6	69.3	61.7	55.3	51.7	48.8	45.4	43.3
80	116.3	112.3	106.6	101.9	96.6	88.1	79.3	71.1	64.3	60.4	57.2	53.5	51.2
90	128.3	124.1	118.1	113.1	107.6	98.6	89.3	80.6	73.3	69.1	65.6	61.8	59.2
100	140.2	135.8	129.6	124.3	118.5	109.1	99.3	90.1	82.4	77.9	74.2	70.1	67.3

Source: Catherine M. Thompson, Table of percentage points of the χ^2 distribution, *Biometrika* **32** (1941), by permission of the author and publisher.

Bibliography

[1] Backhaus J O (1982) A semi-implicit scheme for the shallow water equations for application to shelf sea modelling. *Cont. Shelf Res.* **2**(4), pp243-54.

[2] Baines P.G. (1995) *Topographic Effects in Stratified Flows* C.U.P. 482pp.

[3] Banks J.E. (1974) A mathematical model of a river-shallow sea system used to investigate tide, surge and their interaction in the Thames-southern North Sea region *Phil. Trans. Roy. Soc. A* **275** pp 567 - 609.

[4] Batchelor (1953) *The Theory of Homogeneous Turbulence.* C.U.P.

[5] Berry J S (1996) *Nonlinear Dynamics*, Edward Arnold.

[6] Blumberg A.F. and Mellor G.L. (1987) A description of a three-dimensional coastal ocean circulation model. In: *Three-Dimensional Coastal Ocean Models*, N.Heaps (ed), Coastal and Estuarine Sciences, Series 4, A.G.U. pp1 - 16.

[7] Brink, K.H. (1999) Island-trapped waves, with applications to observations off Bermuda. *Dynamics of Oceans and Atmospheres* 29, 93-118

[8] Buchwald V.T. and Adams J.K. (1968) The propagation of continental shelf waves. *Proc. Roy. Soc. Lond. Ser. A* **305**, pp 235 - 250.

[9] Chatwin P.C. and Allen C.M. (1985) Mathematical models of dispersion in rivers and estuaries *Ann. Rev. Fluid Mech.* **17** pp 119 - 149.

[10] Craik A.D.D. and Leibovich S. (1976) A rational model for Langmuir Circulations, *J. Fluid Mech.* **29** pp 337 - 347.

[11] Crank J. (1998) *The Mathematics of Diffusion.* Oxford Science Pbl. 414 pp.

[12] Csanady G.T. (1997) On the theories that underlie our understanding of continental shelf circulation. *J. Oceanogr.* **53** pp 207 - 229

[13] Dale A.C. and Sherwin T.J. (1996) The extension of baroclinic coastal-trapped wave theory to superinertial frequencies. *J. Phys. Oceanog.* **26**, pp2305-2315.

[14] de Kok J.M. (1994) Tidal averaging and models for anisotropic dispersion in coastal waters *Tellus* **46A** pp 160 - 177

[15] Dingemans M.W. (1997) *Water wave propagation over uneven bottoms, Part 1 Linear Wave Propagation.* World Scientific 471pp

[16] Durran D.R. (1999) *Numerical Methods for Wave Equations in Geophysical Fluid Dynamics.* Springer 465pp.

[17] Dyer K.R. (1997) *Estuaries: A Physical Introduction, (2nd Ed.)* John Wiley 195pp.

[18] Dyke P.P.G. (1977) A simple ocean surface layer model. *Riv. Ital. di Geofisica* **4** pp 31 - 34.

[19] Dyke P.P.G. (1980) On the Stokes' drift induced by tidal motions in a wide estuary *Est. Coast. Mar. Sci.* **11** pp 17 - 25

[20] Dyke P.P.G. (1987) Water circulation in the Firth of Forth, Scotland *Proc. Roy. Soc. Edin.* **93B** pp 273 - 284.

[21] Dyke P.P.G. (1996) *Modelling Marine Processes* Prentice-Hall 158pp.

[22] Dyke P.P.G. and Robertson T. (1985) The simulation of offshore turbulence using seeded eddies. *App. Math. Mod.* **9**(6), pp429-33.

[23] Elliott A J (1991) EUROSPILL: Oceanographic processes and NW European shelf databases. *Mar. Poll. Bull.* **22** pp 548 - 553.

[24] Elliott A J, N Hurford and C J Penn (1986) Shear diffusion and the spreading of oil slicks. *Mar. Poll. Bull.* **17** pp 308 - 313.

[25] Erdogan M.E. and Chatwin P.C. (1967) The effects of curvature and buoyancy on the laminar dispersion of solute in a horizontal tube. *J. Fluid Mech.* **29** pp 465 - 484.

[26] Fischer H.B. (1972) Mass transport mechanisms in partially stratified estuaries, *J. Fluid Mech.* **53** pp 672 - 687

[27] Flather R.A. (1979) In *Marine Forecasting.* Ed. J.C.J. Nihoul (Amsterdam, Elsevier) pp 385-409.

[28] Gill A.E. (1982) *Atmosphere - Ocean Dynamics*, Academic Press 662 pp.

[29] Gill A.E. and Clarke A.J. (1974) Wind induced upwelling, coastal currents and sea-level changes *Deep-Sea Research*, **21**, pp 325 - 345.

[30] Gill A.E. and Schumann A.H. (1974) The generation of long shelf waves by the wind *J. Phys. Oceanog.*, **4** pp 83 - 90.

[31] Givoli D. (1992) *Numerical Methods for Problems in Infinite Domains*, Elsevier.

[32] Heaps N.S. (1971) On the numerical solution of the three dinensional hydrodynamical equations for tides and storm surges. *Mem. Soc Roy Sciences de Liège*, 6e Series, Vol 1 pp 143-180.

[33] Heaps N.S. (1987)(ed) *Three-Dimensional Coastal Ocean Models*, Coastal and Estuarine Sciences 4 AGU, Washington DC, 208pp.

[34] Hinze J.O. (1975) *Turbulence* McGraw-Hill 790pp.

[35] Hirsch C. (1988) *Numerical computation of internal and external flows*, Vol 1. Fundamentals of Numerical Discretisation, John Wiley, 515pp.

[36] Hughes R.L. (1989) The hydraulics of local separation in a coastal current with application to the Kuroshio meander. *J. Phys. Oceanog.* **19** pp 1809 - 1820

[37] Hunter J.R. (1980) An Interactive Computer Model of oil slick motion, *Oceanol. Int . Session M* pp 42-50, Brighton, UK.

[38] Huyer A, R.L.Smith and T.Paluszkiewicz (1987) Coastal upwelling off Peru during normal and El Niño times, 1981 - 1984 *J. Geophys. Res.* **92** pp 14,297 - 14,307.

[39] James I.D. (1996) Advection schemes for shelf sea models. *J. Marine Systems* **8** pp 237 - 254

[40] Klein, P. and Steele, J.H. (1985) Some physical factors affecting ecosystems, *J. Marine Res.* **43**(2), 337-43pp.

[41] LeBlond P.H. and Mysak L.A. (1978) *Waves in the Ocean*. Elsevier Oceanography Series 20, Elsevier Press, 602pp

[42] Lesurier M (1997) *Turbulence in Fluids (3rd Ed)* Kluwer 515pp

[43] Lewis R. (1997) *Dispersion in Estuaries and Coastal Waters*. John Wiley, 312pp.

[44] Longuet-Higgins M.S. (1953) Mass transport in water waves. *Phil. Trans. Roy. Soc. Lond. 345*, pp 535-58

[45] Lynch D.R. and Gray W.G. (1979) A wave equation model for finite element tidal computations. *Computers and Fluids* **7** pp 207-228.

[46] Mellor G.L. (1996) *Introduction to Physical Oceanography* American Institute of Physics, New York, 260pp

[47] Mellor G.L. and Yamada T. (1974) A heirarchy of turbulence closure models for planetary boundary layers. *J. Atmos. Sci.* **31**, 1791-896pp.

[48] Michell A.R. and Wait R. (1985) *Finite Element Analysis and Applications.* (Chichester, Wiley).

[49] Mitsuyasu H., Tasai F., Suhara T., Mizuno S., Ohkusu M., Honda T. and Rikiishi K. (1975) Observations of the directyional spectrum of ocean waves using a cloverleaf buoy. *J. Phys. Oceanog.* **5** pp750 - 760.

[50] Moore D. (1970) The mass transport velocity induced by free oscillations at a single frequency. *Geoph. Fl. Dyn.* **1** pp 237-247.

[51] Murdock J. and Barnes J.A. (1974) *Statistical Tables.*(london and Basingstoke, Macmillan).

[52] Okubo A. (1971) Oceanic diffusion diagrams. *Deep Sea Res.* **18**, pp 781-802.

[53] Okubo A. (1974) Some speculations on oceanic diffusion diagrams. *Rapp.Proc. Verb. Cons. Int. Explor. Mer.* **167**, 77 - 85.

[54] Pond S. and Pickard G.L. (1991) *Introductory Dynamical Oceanography.* 2nd Edn (Oxford, Pergamon).

[55] Proctor Roger, Flather Roger and Elliott Alan J. (1994) Modelling tides and surface drift in the Arabian Gulf - application to the Gulf oil spill *Cont. Shelf Res.*14pp 531 - 545.

[56] Proudman J. (1953) *Dynamical Oceanography.* Methuen 409pp.

[57] Reynolds R.W. (1978) Some Effects of an elliptic reidge on waves of tidal frequency. *J. Phys. Oceanography* **8** pp 38-46.

[58] Roache P.J. (1986) *Computational Fluid Dynamics.* Hermosa.

[59] Robinson A.R. (1964) Contintental shelf waves and the response of sea level to weather systems. *J. Geophys. Res.* **69** pp 367 - 368.

[60] Ruddick K.G., Deleersnijder E., de Mulder T. and Luyten P.J. (1994) A model study of the Rhine discharge *Tellus* **46A** pp 149 - 159

[61] Schlichting H. (1975) *Boundary Layer Theory* (7th Ed.) McGraw Hill 817pp.

[62] Smith G.D. (1965) *Numerical Solutions of Partial Differential Equations.* Oxford University Press. 179pp.

[63] Smith R. (1978) Asymptotic solutions of the Erdogan-Chatwin equation *J. Fluid Mech.* **88** pp 323 - 337.

[64] Smith R. (1979) Buoyancy effects upon lateral dispersion in open channel flow *J. Fluid Mech.* **90** pp 761 - 779.

[65] Smith R. (1980) Buoyancy effects in longitudinal dispersion in wide well-mixed estuaries. *Phil. Trans. Roy. Soc. Lond.* **296A** pp 467 - 496.

[66] Soulsby R. (1997) *Dynamics of Marine Sands. A manual for practical applications.* Thos. Telford 249pp.

[67] Stommel H. (1965) *The Gulf Stream.* 2nd edn (Berkeley, CA, University of California Press).

[68] Summerfield W.C. (1969) *On the trapping of wave energy by bottom topography.* Horace Lamb Centre for Oceanogr. Res. paper 30.

[69] Taylor A.H., Watson A.J., Ainsworth M., Robertson J.E., and Turner D.R. (1991) A modelling investigation of the role of phytoplankton in the surface balance of carbon at the surface of the North Atlantic. *Global Biogeochemical Cycles* 5(2), pp 151-71.

[70] Townsend A.A. (1954) *The Structure of turbulent shear flow* (2nd ed.) 429pp.

[71] Varela R.A., Cruzado A., Tintoré J. and Ladona E.G. (1994) Modelling the deep-chorophyll maximum: A coupled physical-biological approach. *J. Marine Res.* 50(3), pp 441-63.

[72] Watson G.N. (1922) *A treatise on the theory of Bessel functions.* Cambridge University Press, 804pp.

[73] Webb, A.J. and Metcalfe, A.P. (1987) Physical aspects, water movements and modelling studies of the Forth Estuary. *Proc. Roy. Soc. Edin.* **93B** pp 259 - 272

[74] Westerink J.J., Luettich R.A. and Muccino J.C. (1994) Modelling tides in the western North Atlantic using unstructured graded grids. *Tellus* **46A**(2), pp 178-99.

[75] Wolanski E. (1988) Circulation anomalies in tropical Australian estuaries. In: *Hydrodynamics of Estuaries, ed. B. Kjerfve* CRC Press pp 53 - 59.

[76] Wolf J. (1982) A comparison of a semi-implicit with an explicit scheme in a three-dimensional hydrodynamic model. *Cont. Shelf Res.* **2** pp 215-229.

Index

INDEX